ncing learning, changing lives

Decision
Mathematics 2

Edexcel AS and A-level
Modular Mathematics

Susie G Jameson

Contents

About this book

This book is designed to provide you with the best preparation possible for your Edexcel D2 unit examination:

- The LiveText CD-ROM in the back of the book contains even more resources to support you through the unit.

Brief chapter overview and 'links' to underline the importance of mathematics: to the real world, to your study of further units and to your career

Finding your way around the book

Detailed contents list shows which parts of the D2 specification are covered in each section

Every few chapters, a review exercise helps you consolidate your learning

Each section begins with a statement of what is covered in the section

Concise learning points

Step-by-step worked examples

Past examination questions are marked 'E'

Each section ends with an exercise – the questions are carefully graded so they increase in difficulty and gradually bring you up to standard

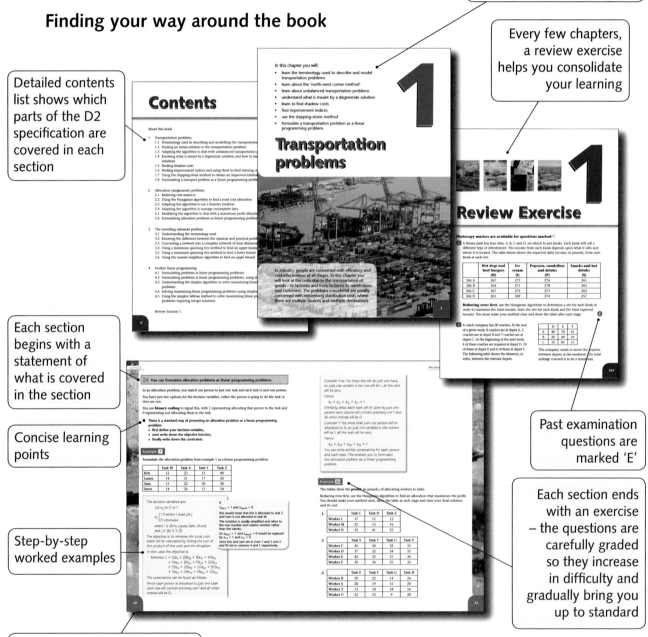

Each chapter has a different colour scheme, to help you find the right chapter quickly

Each chapter ends with a mixed exercise and a summary of key points.

At the end of the book there is an examination-style paper.

LiveText software

The LiveText software gives you additional resources: Solutionbank and Exam café. Simply turn the pages of the electronic book to the page you need, and explore!

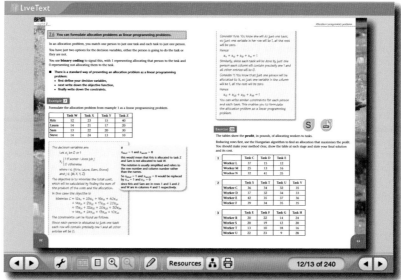

Unique Exam café feature:

- Relax and prepare – revision planner; hints and tips; common mistakes
- Refresh your memory – revision checklist; language of the examination; glossary
- Get the result! – fully worked examination-style paper

Solutionbank

- Hints and solutions to every question in the textbook
- Solutions and commentary for all review exercises and the practice examination paper

Published by Pearson Education Limited, a company incorporated in England and Wales, having its registered office at 80 Strand, London, WC2R 0RL. Registered company number: 872828

Edexcel is a registered trademark of Edexcel Limited

Text © Susie Jameson 2010

17 16
10 9 8 7

British Library Cataloguing in Publication Data
A catalogue record for this book is available from the British Library on request.

ISBN 978 0 435519 20 9

Copyright notice
All rights reserved. No part of this publication may be reproduced in any form or by any means (including photocopying or storing it in any medium by electronic means and whether or not transiently or incidentally to some other use of this publication) without the written permission of the copyright owner, except in accordance with the provisions of the Copyright Designs and Patents Act 1988 or under the terms of a licence issued by the Copyright Licensing Agency, Saffron House, 6–10 Kirby Street, London EC1N 8TS (www.cla.co.uk). Applications for the copyright owner's written permission should be addressed to the publisher.

Edited by Susan Gardner
Typeset by Tech-Set Ltd
Illustrated by Tech-Set Ltd
Cover design by Christopher Howson
Picture research by Chrissie Martin
Cover photo/illustration © Edexcel
Index by Indexing Specialists (UK) Ltd
Printed in Malaysia (CTP-VVP)

Acknowledgements
The author and publisher would like to thank the following individuals and organisations for permission to reproduce photographs:

Shutterstock/Donvictoria@o2.pl p**1**; Shutterstock/Gert Johannes Jacobus Very p**32**; photographersdirect.com/ Wayne Howes Photography p**61**; Digital Vision p**88**; Alamy/Photogenix p**134**; Shutterstock/Ilja Masik p**165**; Rex Features/Action Press p**207**

Every effort has been made to contact copyright holders of material reproduced in this book. Any omissions will be rectified in subsequent printings if notice is given to the publishers.

Disclaimer
This Edexcel publication offers high-quality support for the delivery of Edexcel qualifications.
Edexcel endorsement does not mean that this material is essential to achieve any Edexcel qualification, nor does it mean that this is the only suitable material available to support any Edexcel qualification. No endorsed material will be used verbatim in setting any Edexcel examination/assessment and any resource lists produced by Edexcel shall include this and other appropriate texts.
Copies of official specifications for all Edexcel qualifications may be found on the Edexcel website – www.edexcel.com.

In this chapter you will:

- learn the terminology used to describe and model transportation problems
- learn about the 'north-west corner method'
- learn about unbalanced transportation problems
- understand what is meant by a degenerate solution
- learn to find shadow costs
- find improvement indices
- use the stepping-stone method
- formulate a transportation problem as a linear programming problem.

Transportation problems

In industry, people are concerned with efficiency and cost-effectiveness at all stages. In this chapter you will look at the costs due to the transportation of goods – to factories and from factories to warehouses and customers. The problems considered are usually concerned with minimising distribution costs where there are multiple sources and multiple destinations.

1.1 You should be familiar with the terminology used in describing and modelling the transportation problem.

In order to solve transportation problems you need to consider:

■ The capacity of each of the **supply points** (or sources) – the quantity of goods that can be produced at each factory or held at each warehouse. This is called the **supply** or **stock**.

■ The amount required at each of the **demand points** – the quantity of goods that are needed at each shop or by each customer. This is called the **demand** (or **destination**).

■ The **unit cost** of transporting goods from the supply points to the demand points.

> The **unit cost** is the cost of transporting one item. If one item costs c pounds to transport from A to X then two items will cost $2c$ pounds to transport along that route, and n items nc pounds.

Example 1

Three suppliers A, B and C, each produce road grit which has to be delivered to council depots W, X, Y and Z. The stock held at each supplier and the demand from each depot is known. The cost, in pounds, of transporting one lorry load of grit from each supplier to each depot is also known. This information is given in the table.

	Depot W	Depot X	Depot Y	Depot Z	Stock (lorry loads)
Supplier A	180	110	130	290	14
Supplier B	190	250	150	280	16
Supplier C	240	270	190	120	20
Demand (lorry loads)	11	15	14	10	50

> This table is often referred to as the cost matrix.

> This is the cost of transporting one lorry load from B to Y (in £s).

> Notice that the total supply is equal to the total demand. If this is not the case we simply introduce a dummy destination to absorb the excess supply, with transportation costs all zero (see Section 1.3).

Use the information in the table to write down:

a the number of lorry loads of grit that each supplier can supply

b the number of lorry loads of grit required at each depot

c the cost of transporting a lorry load of grit from A to W

d the cost of transporting a lorry load of grit from C to Z.

e Which is the cheapest route to use?

f Which is the most expensive route to use?

a	Suppliers A, B and C can provide 14, 16 and 20 lorry loads respectively.
b	Depots W, X Y and Z require 11, 15, 14 and 10 lorry loads respectively.
c	The cost of transporting one lorry load from A to W is £180.
d	The cost of transporting one lorry load from C to Z is £120.
e	The cheapest route is A to X at £110 per load.
f	The most expensive route is A to Z at £290 per load

Solving the transportation problem

The method is as follows.

1 First find an initial solution that uses all the stock and meets all the demands.

2 Calculate the total cost of this solution and see if it can be reduced by transporting some goods along a route not currently in the solution.
(If this is not possible then the solution is optimal.)

3 If the cost can be reduced by using a new route, as many units as possible are allocated to this new route to create a new solution.

4 The new solution is checked in the same way as the initial solution to see if it is optimal. If not, any new routes found are included.

5 When no further savings are possible, an optimal solution has been found.

1.2 You can find an initial solution to the transportation problem.

A method often called the '**north-west corner method**' is used.

1 Create a table, with one row for every source and one column for every destination. Each cell represents a route from a source to a destination. Each destination's demand is given at the foot of each column and each source's stock is given at the end of each row. Enter numbers in each cell to show how many units are to be sent along that route.

2 Begin with the top left-hand corner (the north-west corner). Allocate the maximum available quantity to meet the demand at this destination (whilst not exceeding the stock at this source!).

3 As each stock is emptied, move one square down and allocate as many units as possible from the next source until the demand of the destination is met.
As each demand is met, move one square to the right and again allocate as many units as possible.

4 When all the stock is assigned, and all the demands met, stop.

In order to avoid degenerate solutions, movements are made between squares either vertically or horizontally but **never** diagonally. See example 4	For a problem involving m source vertices and n destination vertices you must enter $n + m - 1$ transportation quantities $\geqslant 0$. This will not reduce throughout the problem.	In the examination, problems will be restricted to a maximum of 4 supply points and 4 demand points.

Example 2

	Depot W	Depot X	Depot Y	Depot Z	Stock
Supplier A	180	110	130	290	14
Supplier B	190	250	150	280	16
Supplier C	240	270	190	120	20
Demand	11	15	14	10	50

Use the north-west corner method to find an initial solution to the problem described in example 1 and shown in the table.

1

	W	X	Y	Z	Stock
A					14
B					16
C					20
Demand	11	15	14	10	50

1 Set up the table.

2 Beginning with the north-west corner, start to fill in the number of units you wish to send along each route.

	W	X	Y	Z	Stock
A	11				14
B					16
C					20
Demand	11	15	14	10	50

2 Start to fill in the number of units you wish to send along each route.

Depot W requires 11 lorry loads. This does not exhaust the stock of supplier A.

	W	X	Y	Z	Stock
A	11	3			14
B					16
C					20
Demand	11	15	14	10	50

3 The demand at W has been met so move one square to the right and allocate 14 − 11 = 3 units.

The stock at A is now exhausted. The demand at X has not been met.

	W	X	Y	Z	Stock
A	11	3			14
B		12			16
C					20
Demand	11	15	14	10	50

As the stock at A has been exhausted, move one square down and allocate the maximum possible number of units from supplier B to depot X. In this case, 15 − 3 = 12.

	W	X	Y	Z	Stock
A	11	3			14
B		12	4 •		~~16~~
C					20
Demand	11	15	14	10	50

Now that the demand at X has also been met, move one square to the right and use the remaining stock at B to start to meet the demand at Y.

	W	X	Y	Z	Stock
A	11	3			14
B		12	4		16
C			10 •		20
Demand	11	15	14	10	50

The stock at B is now exhausted (12 + 4 = 16) and so move one square down and use the stock at C to fulfill the remaining demand at Y.

	W	X	Y	Z	Stock
A	11	3			14
B		12	4		16
C			10	10 •	20
Demand	11	15	14	10	50

Finally, move one square to the right and use the remaining stock at C to meet the demand at Z.

This is the final table. All of the stock has been used and all of the demands met.

> The number of occupied cells (routes used) in the table = number of supply points + number of demand points −1.
>
> In this case the
>
> number of occupied cells (routes used) = 6,
>
> number of supply points = 3,
>
> number of demand points = 4
>
> and 6 = 3 + 4 − 1.

Use this table, together with the table showing costs, to work out the total cost of the solution.

	W	X	Y	Z	Stock
A	180	110	130	290	14
B	190	250	150	280	16
C	240	270	190	120	20
Demand	11	15	14	10	50

The cells shaded indicate the costs of the routes currently being used. It is these, together with the number of units being transported along each route, that give the cost of the solution.

The total cost of this solution is

$(11 \times 180) + (3 \times 110) + (12 \times 250) + (4 \times 150)$
$+ (10 \times 190) + (10 \times 120) = £9\,010.$

1.3 You can adapt the algorithm to deal with unbalanced transportation problems.

- When the total supply > total demand, we say the problem is **unbalanced**.

- If the problem is unbalanced we simply add a dummy demand point with a demand chosen so that total supply = total demand, with transportation costs of zero.

Example 3

	A	B	C	Supply
X	9	11	10	40
Y	10	8	12	60
Z	12	7	8	50
Demand	50	40	30	

Three outlets A, B and C are supplied by three suppliers X, Y and Z. The table shows the cost, in pounds, of transporting each unit, the number of units required at each outlet and the number of units available at each supplier.

a Explain why it is necessary to add a dummy demand point in order to solve this problem.

b Add a dummy demand point and appropriate costs to the table.

c Use the north-west corner method to obtain an initial solution.

a The total supply is 150, but the total demand is 120. A dummy is needed to absorb this excess, so that total supply equals total demand.

b

	A	B	C	D	Supply
X	9	11	10	0	40
Y	10	8	12	0	60
Z	12	7	8	0	50
Demand	50	40	30	30	150

We add a dummy column, D, where the demand is 30 (the amount by which the supply exceeds the demand), and the transportation costs are zero (since there is no actual transporting done)! The problem is now **balanced**, the total supply = the total demand.

c

	A	B	C	D	Supply
X	40				40
Y	10	40	10		60
Z			20	30	50
Demand	50	40	30	30	150

1.4 You understand what is meant by a degenerate solution and know how to manage such solutions.

■ In a feasible solution to a transportation problem with m rows and n columns, if the number of cells used is less than $n + m - 1$, then the solution is **degenerate**.

■ This will happen when an entry, other than the last, is made that satisfies the supply for a given row, and at the same time, satisfies the demand for a given column.

■ The algorithm requires that $n + m - 1$ cells are used in every solution, so a zero needs to be placed in a currently unused cell.

Example 4

	A	B	C	Supply
W	10	11	6	30
X	4	5	9	20
Y	3	8	7	35
Z	11	10	9	35
Demand	30	40	50	120

a Demonstrate that the north-west corner method gives a degenerate solution and explain why it is degenerate.

b Adapt your solution to give a non-degenerate initial solution and state its cost.

a

	A	B	C	Supply
W	30			30
X		20		20
Y		20	15	35
Z			35	35
Demand	30	40	50	120

Notice that there has been a diagonal 'move' from cell WA to cell XB. Degenerate solutions can be avoided by not allowing diagonal moves.

This solution is degenerate since it fulfils all the supply and demand needs but only uses 5 cells WA, XB, YB, YC and ZC. There are 4 rows and 3 columns so a non-degenerate solution will use $4 + 3 - 1 = 6$ cells.

b Start by placing the largest possible number in the north-west corner

	A	B	C	Supply
W	30			30
X				20
Y				35
Z				35
Demand	30	40	50	120

Having placed the 30 in the NW corner, you now need to place the next number in the square to its right or in the square underneath.

Since both the supply and the demand are satisfied, this means that you will have to place a zero in either the cell to its right or the cell underneath.

There are two possible initial solutions, depending on where you chose to place the zero.

Either

	A	B	C	Supply
W	30	0		30
X		20		20
Y		20	15	35
Z			35	35
Demand	30	40	50	120

Or

	A	B	C	Supply
W	30			30
X	0	20		20
Y		20	15	35
Z			35	35
Demand	30	40	50	120

In fact the zero can be placed anywhere in the table, but it is convenient to 'stick to the rule' about restricting the movement to one square down or one square right.

Both have a cost of $(30 \times 10) + 0 + (20 \times 5) + (20 \times 8) + (15 \times 7) + (35 \times 9) = £980$

Exercise 1A

Photocopy masters are available on the CD-ROM for the questions in this exercise.

In Questions 1 to 4, the tables show the unit costs of transporting goods from supply points to demand points. In each case:

a use the north-west corner method to find the initial solution,

b verify that, for each solution, the number of occupied cells = number of supply points + number of demand points − 1.

c determine the cost of each initial solution.

1

	P	Q	R	Supply
A	150	213	222	32
B	175	204	218	44
C	188	198	246	34
Demand	28	45	37	110

2

	P	Q	R	S	Supply
A	27	33	34	41	54
B	31	29	37	30	67
C	40	32	28	35	29
Demand	21	32	51	46	150

3

	P	Q	R	Supply
A	17	24	19	123
B	15	21	25	143
C	19	22	18	84
D	20	27	16	150
Demand	200	100	200	500

4

	P	Q	R	S	Supply
A	56	86	80	61	134
B	59	76	78	65	203
C	62	70	57	67	176
D	60	68	75	71	187
Demand	175	175	175	175	700

5

	A	B	C	D	Supply
X	27	33	34	41	60
Y	31	29	37	30	60
Z	40	32	28	35	80
Demand	40	70	50	20	

Four sandwich shops A, B, C and D can be supplied with bread from three bakeries, X, Y, and Z. The table shows the cost, in pence, of transporting one tray of bread from each supplier to each shop, the number of trays of bread required by each shop and the number of trays of bread that can be supplied by each bakery.

a Explain why it is necessary to add a dummy demand point in order to solve this problem, and what this dummy point means in practical terms.

b Use the north-west corner method to determine an initial solution to this problem and the cost of this solution.

6

	K	L	M	N	Supply
A	35	46	62	80	20
B	24	53	73	52	15
C	67	61	50	65	20
D	92	81	41	42	20
Demand	25	10	18	22	

A company needs to supply ready-mixed concrete from four depots A, B, C and D to four work sites K, L, M and N. The number of loads that can be supplied from each depot and the number of loads required at each site are shown in the table above, as well as the transportation cost per load from each depot to each work site.

a Explain what is meant by a degenerate solution.

b Demonstrate that the north-west corner method gives a degenerate solution.

c Adapt your solution to give a non-degenerate intial solution.

7

	L	M	N	Supply
P	3	5	9	22
Q	4	3	7	a
R	6	4	8	11
S	8	2	5	b
Demand	15	17	20	

The table shows a balanced transportation problem. The initial solution, given by the north-west corner method, is degenerate.

a Use this information to determine the values of *a* and *b*.

b Hence write down the initial, degenerate solution given by the north west-corner method.

Finding an improved solution

■ To find an improved solution, you need to:

  1 use the non-empty cells to find the **shadow costs** (see Section 1.5)

  2 use the shadow costs and the empty cells to find **improvement indices** (see Section 1.6)

  3 use the improvement indices and the stepping stone algorithm to find an **improved solution** (see Section 1.7).

1.5 You can find shadow costs.

■ Transportation costs are made up of two components, one associated with the source and one with the destination. These costs of using that route, are called **shadow costs**.

	Depot W	Depot X	Depot Y	Depot Z	Stock
Supplier A	180	110			14
Supplier B		250	150		16
Supplier C			190	120	20
Demand	11	15	14	10	50

In example 2, the cost of £250 in transporting one unit from supplier B to depot X must be dependent on the features – location, toll costs etc, of both B and X.

Using the routes currently in use you can build up equations, showing the cost of transporting one unit, such as

  S(A) + D(X) = 110   and   S(C) + D(Z) = 120   etc.

where S(A), D(X) are the costs due to supply point A and demand point X and so on, respectively.

You need a value for each of the source components and each of the destination components. You do not have sufficient equations for a solution (five equations and six unknowns) but relative costs will do.

> You are only looking at the costs of the routes used in your current solution.

To find the shadow costs, follow these steps.

1 Start with the north-west corner, set the cost linked with its source to zero.

> Put the first source component to zero, and set the destination component to carry the cost of the transportation in row 1.

2 Move along the row to any other non-empty squares and find any other destination costs in the same way.

3 When all possible destination costs for that row have been established, go to the start of the next row.

4 Move along this row to any non-empty squares and use the destination costs found earlier, to establish the source cost for the row. Once that has been done, find any further unknown destination costs.

5 Repeat steps 3 and 4 until all source and destination costs have been found.

Example 5

	Depot W	Depot X	Depot Y	Depot Z	Stock
Supplier A	180	110	130	290	14
Supplier B	190	250	150	280	16
Supplier C	240	270	190	120	20
Demand	11	15	14	10	50

Calculate the shadow costs given by the initial solution of the problem given in example 2 and shown in the table.

Initial solution (see page 5) was

	W	X	Y	Z	Stock
A	11	3			14
B		12	4		16
C			10	10	20
Demand	11	15	14	10	50

Focus on the **costs** of the routes being used — **the non-empty squares**

> Remember to use the cost values not the number of items currently being transported along that route.

	Depot W	Depot X	Depot Y	Depot Z	Stock
Supplier A	180	110			14
Supplier B		250	150		16
Supplier C			190	120	20
Demand	11	15	14	10	50

Putting S(A) to zero, from row 1 we get D(W) = 180
and D(X) = 110

Put S(A) = 0 arbitrarily and
then solve the equations
S(A) + D(W) = 180 and
S(A) + D(X) = 110.

Find the remaining shadow costs by 'walking round' the current
solution, noting the shadow costs you find round the edge
of the table, and using shadow costs found earlier to find the
remaining ones.

Shadow costs		180	110			
		Depot W	Depot X	Depot Y	Depot Z	Stock
0	Supplier A	180	110			14
	Supplier B		250	150		16
	Supplier C			190	120	20
	Demand	11	15	14	10	50

Now move to Row 2.
You know that D(X) = 110, so you find S(B) = 140
hence D(Y) = 10

Knowing that D(X) = 110, you solve
S(B) + D(X) = 250 to get S(B) = 140,
and use this together with the equation
S(B) + D(Y) = 150 to get D(Y) = 10.

Shadow costs		180	110	10		
		Depot W	Depot X	Depot Y	Depot Z	Stock
0	Supplier A	180	110			14
140	Supplier B		250	150		16
	Supplier C			190	120	20
	Demand	11	15	14	10	50

Move to Row 3.
You know that D(Y) = 10, so we find
S(C) = 180 and hence that D(Z) = −60.

Knowing that D(Y) = 10, you solve
S(C) + D(Y) = 190 to get S(C) = 180, and
then solve S(C) + D(Z) = 120 to find D(Z).

Shadow costs		180	110	10	−60	
		Depot W	Depot X	Depot Y	Depot Z	Stock
0	Supplier A	180	110			14
140	Supplier B		250	150		16
180	Supplier C			190	120	20
	Demand	11	15	14	10	50

You have now found all source and all destination shadow costs.
S(A) = 0   S(B) = 140   S(C) = 180
D(W) = 180   D(X) = 110   D(Y) = 10
D(Z) = −60

Do not be alarmed at the negative shadow
cost found for D(Z). It is a feature of
arbitrarily putting S(A) = 0. You are simply
finding costs relative to the cost S(A).

Example 6

	A	B	C	D	Supply
X	9	11	10	0	40
Y	10	8	12	0	60
Z	12	7	8	0	50
Demand	50	40	30	30	150

Calculate the shadow costs given by the initial solution of the problem given in example 3 and shown in the table.

The north-west corner method gave the following initial solution (see page 6)

	A	B	C	D	Supply
X	40				40
Y	10	40	10		60
Z			20	30	50
Demand	50	40	30	30	

We need to use the costs rather than the number of items being transported. So we use the following numbers.

Shadow costs							
		A	B	C	D	Supply
	X	9				40
	Y	10	8	12		60
	Z			8	0	50
	Demand	50	40	30	30	150

Arbitrarily assign S(X) = 0.

Shadow costs							
		A	B	C	D	Supply
0	X	9				40
	Y	10	8	12		60
	Z			8	0	50
	Demand	50	40	30	30	150

Use this to work out the shadow cost for D(A).

Shadow costs		9				
		A	B	C	D	Supply
0	X	9				40
	Y	10	8	12		60
	Z			8	0	50
	Demand	50	40	30	30	150

Use this to work out the shadow cost for S(Y).

Shadow costs		9				
		A	B	C	D	Supply
0	X	9				40
1	Y	10	8	12		60
	Z			8	0	50
	Demand	50	40	30	30	150

We use this to work out the shadow costs for D(B) and D(C).

Shadow costs		9	7	11		
		A	B	C	D	Supply
0	X	9				40
1	Y	10	8	12		60
	Z			8	0	50
	Demand	50	40	30	30	150

Use these to work out the shadow cost for S(Z).

Shadow costs		9	7	11		
		A	B	C	D	Supply
0	X	9				40
1	Y	10	8	12		60
−3	Z			8	0	50
	Demand	50	40	30	30	150

Use this to work out the shadow cost for D(D).

Shadow costs		9	7	11	3	
		A	B	C	D	Supply
0	X	9				40
1	Y	10	8	12		60
−3	Z			8	0	50
	Demand	50	40	30	30	150

You do not have to show each stage of the table in the examination. Just this final list of shadow costs is sufficient.

1.6 You can find improvement indices and use these to find entering cells.

It may be possible to reduce the cost of the initial solution by introducing a route that is not currently in use. You consider each unused route in turn and calculate the reduction in cost which would be made by sending one unit along that route. This is called the **improvement index**.

■ The improvement index in sending a unit from a source P to a demand point Q is found by subtracting the source cost S(P) and destination cost D(Q) from the stated cost of transporting one unit along that route C(PQ). i.e.

$$\text{Improvement index for PQ} = I_{PQ} = C(PQ) - S(P) - D(Q)$$

■ The route with the most negative improvement index will be introduced into the solution.

■ The cell corresponding to the value with the most negative improvement index becomes the **entering cell** (or **entering square** or **entering route**) and the route it replaces is referred to as the **exiting cell** (or **exiting square** or **exiting route**).

■ If there are two equal potential entering cells you may choose either. Similarly, if there are two equal exiting cells, you may select either

■ If there are no negative improvement indices the solution is **optimal**.

Example 7

Shadow costs		180	110	10	−60	
		Depot W	Depot X	Depot Y	Depot Z	Stock
0	Supplier A	180	110	130	290	14
140	Supplier B	190	250	150	280	16
180	Supplier C	240	270	190	120	20
	Demand	11	15	14	10	50

Use the shadow costs found in example 5, and shown in the table above, to calculate improvement indices, and use these to identify the entering cell.

Focus on the routes not currently being used, BW, CW, CX, AY, AZ and BZ.

You already know that

S(A) = 0   S(B) = 140   S(C) = 180   D(W) = 180   D(X) = 110   D(Y) = 10   D(Z) = −60

Improvement index for BW = I_{BW} = C(BW) − S(B) − D(W) = 190 − 140 − 180 = −130

Improvement index for CW = I_{CW} = 240 − 180 − 180 = −120

Improvement index for CX = I_{CX} = 270 − 180 − 110 = −20

Improvement index for AY = I_{AY} = 130 − 0 − 10 = 120

Improvement index for AZ = I_{AZ} = 290 − 0 − (−60) = 350

Improvement index for BZ = I_{BZ} = 280 − 140 − (−60) = 200

The entering cell is therefore BW, since this is the most negative.

Example 8

	X	Y	Z	Supply
A	11	12	17	11
B	13	10	13	15
C	15	18	9	14
Demand	10	15	15	

a Use the north-west corner method to find an initial solution to the transportation problem shown in the table.

b Find the shadow costs and improvement indices.

c Hence determine if the solution is optimal.

a

	X	Y	Z	Supply
A	10	1		11
B		14	1	15
C			14	14
Demand	10	15	15	

b

Shadow costs			11	12	15	
			X	Y	Z	Supply
0		A	11	12	17	11
−2		B	13	10	13	15
−6		C	15	18	9	14
		Demand	10	15	15	

Improvement indices for cells:

BX = 13 + 2 − 11 = 4
CX = 15 + 6 − 11 = 10
CY = 18 + 6 − 12 = 12
AZ = 17 − 0 − 15 = 2

c There are no negative improvement indices, so the solution is optimal.

Exercise 1B

Photocopy masters are available on the CD-ROM for the questions in this exercise.

Questions 1 to 4

Start with the initial, north-west corner, solutions found in questions 1 to 4 of exercise 1A.
In each case use the initial solution, and the original cost matrix, shown below, to find
a the shadow costs,
b the improvement indices
c the entering cell, if appropriate.

1

	P	Q	R	Supply
A	150	213	222	32
B	175	204	218	44
C	188	198	246	34
Demand	28	45	37	

2

	P	Q	R	S	Supply
A	27	33	34	41	54
B	31	29	37	30	67
C	40	32	28	35	29
Demand	21	32	51	46	

3

	P	Q	R	Supply
A	17	24	19	123
B	15	21	25	143
C	19	22	18	84
D	20	27	16	150
Demand	200	100	200	

4

	P	Q	R	S	Supply
A	56	86	80	61	134
B	59	76	78	65	203
C	62	70	57	67	176
D	60	68	75	71	187
Demand	175	175	175	175	

1.7 You can use the stepping-stone method to obtain an improved solution.

In example 7 you discovered that the most negative improvement index was BW with a value of −130 (see page 15). This means that every time you send a unit along BW you save a cost of 130. Therefore you want to send as many units as possible along this new route. You have to be careful, however, not to exceed the stock or the demand. To ensure this, you go through a sequence of adjustments, called the **stepping-stone method**.

You are looking therefore for a cycle of adjustments, where you increase the value in one cell and then decrease the value in the next cell, then increase the value in the next, and so on.

A popular mind picture is that you are using the cells as 'stepping-stones', placing one foot on each, and alternately putting down your left foot (increasing) then right foot (decreasing) as you journey around the table – hence the method's nickname.

The stepping-stone method

1 Create the cycle of adjustments. The two basic rules are:

    a within any row and any column there can only be one increasing cell and one decreasing cell.

    b apart from the entering cell, adjustments are only made to non-empty cells.

2 Once the cycle of adjustments has been found you transfer the maximum number of units through this cycle. This will be equal to the smallest number in the decreasing cells (since you may not have negative units being transported).

3 You then adjust the solution to incorporate this improvement.

Example 9

In example 2, the table of costs was

	Depot W	Depot X	Depot Y	Depot Z	Stock
Supplier A	180	110	130	290	14
Supplier B	190	250	150	280	16
Supplier C	240	270	190	120	20
Demand	11	15	14	10	50

and the initial solution was

	W	X	Y	Z	Stock
A	11	3			14
B		12	4		16
C			10	10	20
Demand	11	15	14	10	50

at a cost of £9 010 (see page 5).

Obtain an improved solution and find the improved cost.

Use BW as the entering cell, since this gave the most negative improvement index, −130 (see Example 7, page 15). So BW will be an increasing cell. We enter a value of θ into this cell.

	W	X	Y	Z	Stock
A	11	3			14
B	θ	12	4		16
C			10	10	20
Demand	11	15	14	10	50

In order to keep the demand at W correct, you must therefore decrease the entry at AW, so AW will be a decreasing cell.

	W	X	Y	Z	Stock
A	$11 - \theta$	3			14
B	θ	12	4		16
C			10	10	20
Demand	11	15	14	10	50

In order to keep the stock at A correct, you must therefore increase the entry at AX, so AX will be an increasing cell.

	W	X	Y	Z	Stock
A	$11 - \theta$	$3 + \theta$			14
B	θ	12	4		16
C			10	10	20
Demand	11	15	14	10	50

In order to keep the demand at X correct, you must therefore decrease the entry at BX, so BX will be a decreasing cell.

	W	X	Y	Z	Stock
A	$11 - \theta$	$3 + \theta$			14
B	θ	$12 - \theta$	4		16
C			10	10	20
Demand	11	15	14	10	50

This is as far as you can go with adjustments since the top two rows both have an increasing and decreasing cell.

Now choose a value for θ, the greatest value you can, without introducing negative entries into the table. Look at the decreasing cells and see that the greatest value of θ is 11 (since $11 - 11 = 0$).

Replace θ by 11 in the table:

	W	X	Y	Z	Stock
A	$11 - 11$	$3 + 11$			14
B	11	$12 - 11$	4		16
C			10	10	20
Demand	11	15	14	10	50

This gives the improved solution:

	W	X	Y	Z	Stock
A		14			14
B	11	1	4		16
C			10	10	20
Demand	11	15	14	10	50

Looking at the table of transportation costs, this solution has a cost of £7 580

As a double check, it is always true that

New cost = cost of former solution + improvement index × θ.

In this case $7\,580 = 9\,010 + (-130) \times 11$

You will notice that AW has become empty. AW is therefore the **exiting cell.**

Remember that the number of cells used in a feasible solution must equal the number of rows plus the number of columns minus 1. So if you put a number into an entering cell, it must be balanced by a number being removed from an exiting cell.

■ **At each iteration we create one entering cell and one exiting cell.**

■ **To find an optimal solution, continue to calculate new shadow costs and improvement indices and then apply the stepping-stone method. Repeat this iteration until all the improvement indices are non-negative.**

Example 10

Find an optimal solution for example 9.

This second iteration indicates the amount of working you need to show in the examination.

Second iteration

Find the new shadow costs:

Shadow costs		50	110	10	−60	
		W	X	Y	Z	Stock
0	A	180	110	130	290	14
140	B	190	250	150	280	16
180	C	240	270	190	120	20
	Demand	11	15	14	10	50

Finding the new improvement indices for the non-used cells:

$AW = 180 - 0 - 50 = 130$

$CW = 240 - 180 - 50 = 10$

$CX = 270 - 180 - 110 = -20$

$AY = 130 - 0 - 10 = 120$

$AZ = 290 - 0 + 60 = 350$

$BZ = 280 - 140 + 60 = 200$

So the new entering cell is CX, since this has the most negative improvement index.

Applying the stepping-stone method gives

	W	X	Y	Z	Stock
A		14			14
B	11	$1 - \theta$	$4 + \theta$		16
C		θ	$10 - \theta$	10	20
Demand	11	15	14	10	50

Looking at cells BX and CY we see that the greatest value for θ is 1

The new exiting cell will be BX, $\theta = 1$ and we get

	W	X	Y	Z	Stock
A		14			14
B	11		5		16
C		1	9	10	20
Demand	11	15	14	10	50

The new cost is £7 560

Checking, $7\,580 + (-20) \times 1 = 7560$

Third iteration

New shadow costs

Shadow costs		70	110	30	−40	
		W	X	Y	Z	Stock
0	A	180	110	130	290	14
120	B	190	250	150	280	16
160	C	240	270	190	120	20
	Demand	11	15	14	10	50

New improvement indices for the non-used cells:

AW = 180 − 0 − 70 = 110
CW = 240 − 160 − 70 = 10
BX = 250 − 120 − 110 = 20
AY = 130 − 0 − 30 = 100
AZ = 290 − 0 + 40 = 330
BZ = 280 − 120 + 40 = 200

There are no negative improvement indices so this solution is optimal.

The solution is   110 units A to X
190 units B to W
150 units B to Y
270 units C to X
190 units C to Y
120 units C to Z

At this point, if there is an improvement index of 0, this would indicate that there is an alternative optimal solution. To find it, simply use the cell with the zero improvement index as the entering cell. (See question 2 in Mixed Exercise 1E.)

Some stepping-stone routes are not rectangles and some θ values are not immediately apparent.

Example 11

	Supermarket X	Supermarket Y	Supermarket Z	Stock
Warehouse A	24	22	28	13
Warehouse B	26	26	14	11
Warehouse C	20	22	20	12
Demand	10	13	13	

The table shows the unit cost, in pounds, of transporting goods from each of three warehouses, A, B and C to each of three supermarkets X, Y and Z. It also shows the stock at each warehouse and the demand at each supermarket.

Solve the transportation problem shown in the table. Use the north-west corner method to obtain an initial solution. You must state your shadow costs, improvement indices, stepping-stone routes, θ values, entering cells and exiting cells. You must state the initial cost and the improved cost after each iteration.

Check the problem is balanced, Supply = Demand = 36, so we do not need to add a dummy.
The north-west corner method gives the following initial solution

	X	Y	Z	Stock
A	10	3		13
B		10	1	11
C			12	12
Demand	10	13	13	

The cost of the initial solution is £820

Calculate shadow costs

Shadow costs		24	22	10	
		X	Y	Z	Stock
0	A	24	22	28	13
4	B	26	26	14	11
10	C	20	22	20	12
	Demand	10	13	13	

Calculating the improvement indices for the empty cells:

$BX = 26 - 4 - 24 = -2$
$CX = 20 - 10 - 24 = -14$
$CY = 22 - 10 - 22 = -10$
$AZ = 28 - 0 - 10 = 18$

Use CX as the entering cell, since it has the most negative improvement index.

	X	Y	Z	Stock
A	$10 - \theta$	$3 + \theta$		13
B		$10 - \theta$	$1 + \theta$	11
C	θ		$12 - \theta$	12
Demand	10	13	13	

> This stepping stone route is quite complicated. Take a minute or two to check how it has been created. Start by putting θ in CX, then to correct the demand of X, subtract θ from cell AX, then to correct the supply in A, add θ to cell AY and so on, finishing at cell CZ.

The maximum value of θ is 10

Either cell AX or cell BY can be the exiting cell, since both of these will go to zero. We simply choose one of them, AX to be the exiting cell, the other will have a numerical value of zero.

> Many candidates fail to understand the difference between an empty cell and one with a zero entry. Zero is a number, just like 4 and 'counts' towards our $m + n - 1$ entries in the table. An empty cell has no number in it.

Improved solution is

	X	Y	Z	Stock
A		13		13
B		0	11	11
C	10		2	12
Demand	10	13	13	

The cost is now £680

We need to check for optimality, by calculating improvement indices.

The second set of shadow costs are:

Shadow costs		10	22	10	
		X	Y	Z	Stock
0	A	24	22	28	13
4	B	26	26	14	11
10	C	20	22	20	12
	Demand	10	13	13	

Improvement indices

AX	BX	CY	AZ
14	12	−10	18

So the solution is not yet optimal and the next entering cell is CY.

	X	Y	Z	Stock
A		13		13
B		$0 - \theta$	$11 + \theta$	11
C	10	θ	$2 - \theta$	12
Demand	10	13	13	

We can see from cell BY that the maximum value of θ is 0.

The entering cell is CY and will have an entry of 0. The exiting cell is BY and it will now be empty.

> This looks odd, but the algorithm must be followed. θ can not be negative but it can be any non-negative number, and, as has been already mentioned, 0 is a number.

The improved solution is:

	X	Y	Z	Stock
A		13		13
B			11	11
C	10	0	2	12
Demand	10	13	13	

The cost is unchanged at £680

We again need to check for optimality, by calculating improvement indices.

Shadow costs:

Shadow costs		20	22	20	
		X	Y	Z	Stock
0	A	24	22	28	13
−6	B	26	26	14	11
0	C	20	22	20	12
	Demand	10	13	13	

Improvement indices are

AX	BX	BY	AZ
4	12	10	8

All the improvement indices are non-negative and so the solution is optimal.

	X	Y	Z	Stock
A		13		13
B			11	11
C	10	0	2	12
Demand	10	13	13	

So the optimal solution is to send 13 units from A to Y

11 units from B to Z

10 units from C to X

2 units from C to Z

at a cost of £680

Exercise 1C

Questions 1 to 3

Complete your solutions to the transportation problems from questions 1, 2 and 4 in exercises 1A and 1B. You should demonstrate that your solution is optimal.

1

	P	Q	R	Supply
A	150	213	222	32
B	175	204	218	44
C	188	198	246	34
Demand	28	45	37	

2

	P	Q	R	S	Supply
A	27	33	34	41	54
B	31	29	37	30	67
C	40	32	28	35	29
Demand	21	32	51	46	

3

	P	Q	R	S	Supply
A	56	86	80	61	134
B	59	76	78	65	203
C	62	70	57	67	176
D	60	68	75	71	187
Demand	175	175	175	175	

The solution to question **3** requires a number of iterations, plus the optimality check – you will certainly get lots of practise in implementing the algorithms!

4

	P	Q	Stock
A	2	6	3
B	2	7	5
C	6	9	2
Demand	6	4	

The table shows the unit cost, in pounds, of transporting goods from each of three warehouses A, B and C to each of two supermarkets P and Q. It also shows the stock at each warehouse and the demand at each supermarket.

Solve the transportation problem shown in the table. Use the north-west corner method to obtain an initial solution. You must state your shadow costs, improvement indices, stepping-stone routes, θ values, entering cells and exiting cells. You must state the initial cost and the improved cost after each iteration.

1.8 You can formulate a transportation problem as a linear programming problem.

In book D1 you met linear programming problems in two variables. In chapter 4 of this book you will study linear programming problems in more than two variables.

Consider our first example

	Depot W	Depot X	Depot Y	Depot Z	Stock
Supplier A	180	110	130	290	14
Supplier B	190	250	150	280	16
Supplier C	240	270	190	120	20
Demand	11	15	14	10	50

Let x_{11} (the entry in the 1st row, 1st column) be the number of units transported from A to W, and x_{24} (the entry in the 2nd row and 4th column) be the number of units transported from B to Z, and so on, then you have the following solution.

x_{11}, x_{23}, x_{34} and so on are called the **decision variables**.

	Depot W	Depot X	Depot Y	Depot Z	Stock
Supplier A	x_{11}	x_{12}	x_{13}	x_{14}	14
Supplier B	x_{21}	x_{22}	x_{23}	x_{24}	16
Supplier C	x_{31}	x_{32}	x_{33}	x_{34}	20
Demand	11	15	14	10	50

In this case a lot of these entries will be empty – there will be only $3 + 4 - 1 = 6$ non-empty cells, and all the rest will be blank. However, since you do not yet know which will be empty you allow for any of them to be non-empty as you formulate the problem.

The objective is to minimise the total cost, which will be calculated by finding the sum of the product of number of units transported along each route and the cost of using that route.

In this case the objective is

$$\begin{aligned} \text{Minimise } C = {}& 180x_{11} + 110x_{12} + 130x_{13} + 290x_{14} \\ &+ 190x_{21} + 250x_{22} + 150x_{23} + 280x_{24} \\ &+ 240x_{31} + 270x_{32} + 190x_{33} + 120x_{34} \end{aligned}$$

Looking at the stock, you know that the four entries in the first row must not exceed 14, so you have our first constraint:

$$x_{11} + x_{12} + x_{13} + x_{14} \leqslant 14$$

We can write similar constraints for each supplier and each depot, and include non-negativity constraints. This enables you to formulate the transportation problem as a linear programming problem.

■ **The standard way of presenting a transportation problem as a linear programming problem:**
  - **first define your decision variables,**
  - **next write down the objective function,**
  - **finally write down the constraints.**

Example 12

	R	S	T	Supply
A	3	3	2	25
B	4	2	3	40
C	3	4	3	31
Demand	30	30	36	

Formulate the transportation problem as a linear programming problem. You must state your decision variables, objective and constraints.

Let x_{ij} be the number of units transported from i to j.

    where $i \in \{A, B, C\}$

          $j \in \{R, S, T\}$

   and $x_{ij} \geqslant 0$

FIRST define your decision variables

$$\begin{aligned} \text{Minimise } C = {}& 3x_{11} + 3x_{12} + 2x_{13} \\ &+ 4x_{21} + 2x_{22} + 3x_{23} \\ &+ 3x_{31} + 4x_{32} + 3x_{33} \end{aligned}$$

NEXT write down the objective function.

Subject to:
$$\begin{aligned} x_{11} + x_{12} + x_{13} &\leqslant 25 \\ x_{21} + x_{22} + x_{23} &\leqslant 40 \\ x_{31} + x_{32} + x_{33} &\leqslant 31 \\ x_{11} + x_{21} + x_{31} &\leqslant 30 \\ x_{12} + x_{22} + x_{32} &\leqslant 30 \\ x_{13} + x_{23} + x_{33} &\leqslant 36 \end{aligned}$$

FINALLY write down the constraints.

Exercise 1D

Formulate the following transportation problems as linear programming problems.

1

	P	Q	R	Supply
A	150	213	222	32
B	175	204	218	44
C	188	198	246	34
Demand	28	45	37	

2

	P	Q	R	S	Supply
A	27	33	34	41	54
B	31	29	37	30	67
C	40	32	28	35	29
Demand	21	32	51	46	

3

	P	Q	R	Supply
A	17	24	19	123
B	15	21	25	143
C	19	22	18	84
D	20	27	16	150
Demand	200	100	200	

4

	P	Q	R	S	Supply
A	56	86	80	61	134
B	59	76	78	65	203
C	62	70	57	67	176
D	60	68	75	71	187
Demand	175	175	175	175	

Mixed exercise 1E

1

	L	M	Supply
A	20	70	15
B	40	30	5
C	60	90	8
Demand	16	12	

The table shows the cost, in pounds, of transporting a car from each of three factories A, B and C to each of two showrooms L and M. It also shows the number of cars available for delivery at each factory and the number required at each showroom.

a Use the north-west corner method to find an initial solution.

b Solve the transportation problem, stating shadow costs, improvement indices, entering cells, stepping-stone routes, θ values and exiting cells.

c Demonstrate that your solution is optimal and find the cost of your optimal solution.

d Formulate this problem as a linear programming problem, making your decision variables, objective function and constraints clear.

e Verify that your optimal solution lies in the feasible region of the linear programming problem.

2

	P	Q	R	Supply
F	23	21	22	15
G	21	23	24	35
H	22	21	23	10
Demand	10	30	20	

The table shows the cost of transporting one unit of stock from each of three supply points F, G and H to each of three sales points P, Q and R. It also shows the stock held at each supply point and the amount required at each sales point.

a Use the north-west corner method to obtain an initial solution.

b Taking the most negative improvement index to indicate the entering square, perform two complete iterations of the stepping-stone method. You must state your shadow costs, improvement indices, stepping-stone routes and exiting cells.

c Explain how you can tell that your current solution is optimal.

d State the cost of your optimal solution.

e Taking the zero improvement index to indicate the entering square, perform one futher iteration to obtain a second optimal solution.

3

	X	Y	Z	Supply
J	8	5	7	30
K	5	5	9	40
L	7	2	10	50
M	6	3	15	50
Demand	25	45	100	

The transportation problem represented by the table above is to be solved.

A possible north-west corner solution is

	X	Y	Z	Supply
J	25	5		30
K		40		40
L		0	50	50
M			50	50
Demand	25	45	100	

a Explain why it is was necessary to add a zero entry (in cell LY) to the solution.

b State the cost of this initial solution.

c Choosing cell MX as the entering cell, perform one iteration of the stepping-stone method to obtain an improved solution. You must make your route clear, state your exiting cell and the cost of the improved solution.

d Determine whether your current solution is optimal. Give a reason for your answer.

After two more iterations the following solution was found.

	X	Y	Z	Supply
J			30	30
K		20	20	40
L			50	50
M	25	25		50
Demand	25	45	100	

e Taking the most negative improvement index to indicate the entering square, perform one further complete iteration of the stepping-stone method to obtain an optimal solution. You must state your shadow costs, improvement indices, stepping-stone route and exiting cell.

4

	S	T	U	Supply
A	6	10	7	50
B	7	5	8	70
C	6	7	7	50
Demand	100	30	20	

a Explain why a dummy demand point might be needed when solving a transportation problem.

The table shows the cost, in pounds, of transporting one van load of fruit tree seedlings from each of three greenhouses A, B and C to three garden centres S, T and U. It also shows the stock held at each greenhouse and the amount required at each garden centre.

b Use the north-west corner method to obtain an initial solution.

c Taking the most negative improvement index in each case to indicate the entering square, use the stepping-stone method to obtain an optimal solution. You must state your shadow costs, improvement indices, stepping-stone routes, entering squares and exiting cells.

d State the cost of your optimal solution.

e Formulate this problem as a linear programming problem. Make your decision variables, objective function and constraints clear.

Summary of key points

1 In solving transportation problems, you need to know
- the supply or stock
- the demand
- the unit cost of transporting goods.

2 When total supply > total demand, the problem is **unbalanced**.

3 A dummy destination is needed if total supply does not equal total demand. Transport costs to this dummy destination are zero.

4 To solve the transportation problem, the **north-west corner** method is used.

5 The north-west corner method:
- Create a table, with one row for every source and one column for every destination. Each destination's demand is given at the foot of each column and each source's stock is given at the end of each row. Enter numbers in each cell to show how many units are to be sent along that route.
- Begin with the top left-hand corner. Allocate the maximum available quantity to meet the demand at this destination (but do not exceed the stock at this source).
- As each stock is emptied, move one square down and allocate as many units as possible from the next source until the demand of the destination is met.
- As each demand is met, move one square to the right and again allocate as many units as possible.
- Stop when all the stock is assigned and all the demands are met.

6 In a feasible solution to a transportation problem with m rows and n columns, if the number of cells used is less than $n + m - 1$ then the solution is **degenerate**.

7 The algorithm requires $n + m - 1$ cells to be used in every solution, so a zero must be placed in an unused cell in a degenerate solution.

8 To find an improved solution, you need to:
- use the non-empty cells to find the **shadow costs**
- use the shadow costs and the empty cells to find **improvement indices**
- use the improvement indices and the **stepping-stones method** to find an improved solution.

9 Transportation costs are made up of two components, one associated with the source and one with the destination. The costs of using a route are called **shadow costs**. (See example 5 for how to work out shadow costs.)

10 The **improvement index** of a route is the reduction in cost which would be made by sending one unit along that route.
Improvement index for route PQ = $I_{PQ} = C(PQ) - S(P) - D(Q)$ (see example 7).

11 The stepping-stone method is used to find an improved solution (see example 9).

2

In this chapter you will:
- learn to reduce cost matrices
- use the Hungarian algorithm to find a least cost allocation
- adapt the Hungarian algorithm to use a dummy location
- adapt the Hungarian algorithm to manage incomplete data
- modify the Hungarian algorithm to deal with a maximum profit allocation
- formulate allocation problems as linear programming problems.

Allocation (assignment) problems

In a medley relay swimming team, four swimmers will each swim one length, one after the other, and each will swim a different stroke. Although all the team members could swim any of the strokes, some members of the team are faster at one or two particular strokes. We want the strokes to be allocated to the team members in such a way as to minimise the total time it takes the team to complete all four lengths.

This is an example of an allocation (assignment) problem.

It can be seen as a weighted matching problem, where we seek to minimise the total cost or time or to maximise the total profit.

2.1 You should be able to reduce cost matrices.

In allocation problems each worker must do just one task and each task must be done by just one worker. We are seeking a one-to-one solution.

This means that we require the same number of tasks as workers.

In finding a solution to an allocation problem, only **relative** costs are important.

> The 'cost' might be time, money, or some other quantity.

Suppose each member of the swimming team is timed doing a length of crawl and the times are 12, 13, 11 and 14 seconds. The important thing is not how long they each take, but how much quicker or slower each one is relative to the others. We can subtract 11 seconds from each time to compare the swimmers more easily.

To reduce the cost matrix:

■ subtract the least value in each row from each element of that row,

■ using the new matrix, subtract the least value in each column from each element in that column.

Example 1

	Task W	Task X	Task Y	Task Z
Kris	12	23	15	40
Laura	14	21	17	20
Sam	13	22	20	30
Steve	14	24	13	10

The table shows the times taken, in minutes, by four workers to complete each task of a four-stage production process. The time taken for the process is the sum of the times of the four separate tasks. Reduce the cost matrix.

The smallest numbers in rows 1, 2, 3 and 4 are 12, 14, 13 and 10 respectively. If we subtract these numbers from each element in the row our table becomes

	Task W	Task X	Task Y	Task Z
Kris	0	11	3	28
Laura	0	7	3	6
Sam	0	9	7	17
Steve	4	14	3	0

The smallest numbers in columns 1, 2, 3 and 4 are 0, 7, 3 and 0 respectively. If we subtract these numbers form each element in the column our table becomes

	Task W	Task X	Task Y	Task Z
Kris	0	4	0	28
Laura	0	0	0	6
Sam	0	2	4	17
Steve	4	7	0	0

> In the examination you will be required to reduce the rows first and then the columns.

If we can find a matching using only the cells showing a zero cost we will have found an optimal solution. In this case we can.

Our optimal solution is

Kris does task Y   Laura does task X
Sam does task W and Steve does task Z

> If you find it difficult to 'see' the solution from the table, you could draw a bipartite graph showing the 'zero cost' cells as edges and apply the matching algorithm.

If you look at the original cost matrix the total cost is
$15 + 21 + 13 + 10 = 59$.

Check
The total cost was 59 which is also the value of all the row and column reductions we made ($12 + 14 + 13 + 10 + 7 + 3 = 59$).

> There may be more than one optimal solution, of course, but all will have the same 'cost'.

2.2 You can use the Hungarian algorithm to find a least cost allocation.

The Hungarian algorithm

1 Start by finding the reduced cost matrix.

2 Determine the minimum number of straight lines (horizontal or vertical), which will cover all of the zeros in the matrix.

3 In an $n \times n$ matrix, if you cannot cover them in fewer than n lines, you have an optimal solution and you stop.

4 If in an $n \times n$ matrix you can cover the zeros with fewer than n lines, drawn vertically or horizontally, the solution can be improved.

5 Draw in these lines and look for the smallest uncovered element, e.

6 Add e to the elements in each covered row and each covered column, adding it twice to any element covered twice.

7 Subtract e from every element of the matrix.

8 Repeat steps **2–7** until an optimal solution is found.

> By adding e to whole rows and whole columns in step 6, and by subtracting it from each element in the table in step 7 we are still using relative costs.

The idea of this algorithm is rather like making bids at an auction. We 'offer' up an amount of money − the minimum we can sensibly make, made up of the sum of the row and column reductions.

If we can find a solution it will therefore be optimal.

If we do not find a solution, then we increase our 'bid' and see if we can find a solution for that amount; and keep on increasing until we are successful.

Example 2

Boris, Percival and Spike do garden maintenance. The table shows the time, in minutes, that they would take to do each task in Mrs Green's garden. Allocate the tasks so that the time taken to do the whole job is as small as possible.

	Dig vegetable patch	Weed flower beds	Cut lawn and hedges
Boris	250	80	160
Percival	230	90	150
Spike	230	110	140

	Dig vegetable patch	Weed flower beds	Cut lawn and hedges
Boris	170	0	80
Percival	140	0	60
Spike	120	0	30

First we reduce the cost matrix.
Subtracting 80, 90 and 110 from rows 1, 2 and 3 respectively gives:

	Dig vegetable patch	Weed flower beds	Cut lawn and hedges
Boris	50	0	50
Percival	20	0	30
Spike	0	0	0

Subtracting 120, 0 and 30 from columns 1, 2 and 3 respectively gives:

We can cover all the zeros using only 2 lines like this:

	Dig vegetable patch	Weed flower beds	Cut lawn and hedges
Boris	50	0	50
Percival	20	0	30
Spike	0	0	0

We now need to test for optimality. Since this is a 3 × 3 matrix, we will have an optimal solution if we **have** to use 3 lines (drawn vertically or horizontally) to cover all the zeros.

If we can cover them in fewer than 3 lines, the solution is not optimal.

So we have not yet found a solution – there is no way of allocating each worker to a task using just the zero cost cells.

The minimum uncovered element is 20.

We add 20 to each row covered by a line.

	Dig vegetable patch	Weed flower beds	Cut lawn and hedges
Boris	50	0	50
Percival	20	0	30
Spike	20	20	20

and then to each column covered by a line, giving:

	Dig vegetable patch	Weed flower beds	Cut lawn and hedges
Boris	50	20	50
Percival	20	20	30
Spike	20	40	20

We then subtract 20 from each element in the matrix giving:

	Dig vegetable patch	Weed flower beds	Cut lawn and hedges
Boris	30	0	30
Percival	0	0	10
Spike	0	20	0

We have to use three lines to cover all the zeros, so our current solution is optimal.

A minimal matching is

    Boris – weeding flower beds
    Percival – dig vegetable patch
    Spike – cut lawn and hedges

Using the original cost matrix the cost
= 80 + 230 + 140 = 450 minutes

> With only one iteration needed our 'check' still works .
> If we add together the row and column reductions we get
> 80 + 90 + 110 + 120 + 0 + 30 = 430.
> If we now add on the extra 20 used in the single iteration we get
> 430 + 20 = 450.

A Short Cut
Steps 6 and 7 of the algorithm can be done in just one matrix.
If the smallest uncovered element is 'e' the effect of these three operations is that
i   each element covered by two lines will increase by 'e',
ii   each element covered by just one line will be unchanged,
iii each uncovered element will be reduced by 'e'.

Example 3

	Cut	Sew	Fill	Finish	Pack
Ben	9	8	3	6	10
Ellie	5	5	7	5	5
Greg	10	9	3	9	10
Hyo	10	7	2	9	7
Toby	9	8	2	7	10

Five workers Ben, Ellie, Greg, Hyo and Toby are to be assigned to five tasks involved in making a soft toy. The table shows the time in seconds taken to complete each task.

Reducing rows first, use the Hungarian algorithm to determine an allocation that minimises the total time.

> In the examination the size of the cost matrix will be at most 5 × 5.

Reducing rows first

	Cut	Sew	Fill	Finish	Pack
Ben	6	5	0	3	7
Ellie	0	0	2	0	0
Greg	7	6	0	6	7
Hyo	8	5	0	7	5
Toby	7	6	0	5	8

We can not reduce the columns, since there is a zero already in each column.

We can cover all the zeros in just two lines like this:

	Cut	Sew	Fill	Finish	Pack
Ben	6	5	0	3	7
Ellie	0	0	2	0	0
Greg	7	6	0	6	7
Hyo	8	5	0	7	5
Toby	7	6	0	5	8

Method I: Using the short cut

Our matrix is currently

	Cut	Sew	Fill	Finish	Pack
Ben	6	5	0	3	7
Ellie	0	0	2	0	0
Greg	7	6	0	6	7
Hyo	8	5	0	7	5
Toby	7	6	0	5	8

To use the short cut we need to tackle the elements according to how many lines cover them.

The smallest uncovered element is 3, so using the short cut, we
- Add 3 to the element covered by two lines
- **Leave the elements covered by just one line unchanged**
- Subtract 3 from the uncovered elements

This gives the following matrix

	Cut	Sew	Fill	Finish	Pack
Ben	3	2	0	0	4
Ellie	0	0	5	0	0
Greg	4	3	0	3	4
Hyo	5	2	0	4	2
Toby	4	3	0	2	5

Method II: Not using the short cut
- add 3 to each element in the 'Ellie' row

	Cut	Sew	Fill	Finish	Pack
Ben	6	5	0	3	7
Ellie	3	3	5	3	3
Greg	7	6	0	6	7
Hyo	8	5	0	7	5
Toby	7	6	0	5	8

- add 3 to each element in the 'Fill' column

	Cut	Sew	Fill	Finish	Pack
Ben	6	5	3	3	7
Ellie	3	3	8	3	3
Greg	7	6	3	6	7
Hyo	8	5	3	7	5
Toby	7	6	3	5	8

Notice that the element in the 'Ellie' row and the 'Fill' column has been increased by 6.

- We now subtract 3 from all elements in the table giving:

	Cut	Sew	Fill	Finish	Pack
Ben	3	2	0	0	4
Ellie	0	0	5	0	0
Greg	4	3	0	3	4
Hyo	5	2	0	4	2
Toby	4	3	0	2	5

The matrix resulting from both methods can be covered in just three lines like this:

	Cut	Sew	Fill	Finish	Pack
Ben	3	2	0	0	4
Ellie	0	0	5	0	0
Greg	4	3	0	3	4
Hyo	5	2	0	4	2
Toby	4	3	0	2	5

Using the short cut again we categorise the elements according to how many lines cover them.

	Cut	Sew	Fill	Finish	Pack
Ben	3	2	0	0	4
Ellie	0	0	5	0	0
Greg	4	3	0	3	4
Hyo	5	2	0	4	2
Toby	4	3	0	2	5

The minimum uncovered element is 2, so we

- Add 2 to the element covered by two lines
- **Leave the elements covered by just one line unchanged**
- Subtract 2 from the uncovered elements

Giving

	Cut	Sew	Fill	Finish	Pack
Ben	1	0	0	0	2
Ellie	0	0	7	2	0
Greg	2	1	0	3	2
Hyo	3	0	0	4	0
Toby	2	1	0	2	3

This can be covered in four lines, so we have not yet found a solution

The four lines can be placed like this:

	Cut	Sew	Fill	Finish	Pack
Ben	1	0	0	0	2
Ellie	0	0	7	2	0
Greg	2	1	0	3	2
Hyo	3	0	0	4	0
Toby	2	1	0	2	3

Sometimes there might be a choice in how the lines are arranged. If so, any one of these arrangements can be selected. (However, see page 50.)

Using the short cut again

	Cut	Sew	Fill	Finish	Pack
Ben	1	0	0	0	2
Ellie	0	0	7	2	0
Greg	2	1	0	3	2
Hyo	3	0	0	4	0
Toby	2	1	0	2	3

The minimum uncovered element is 1, so we

- Add 1 to the element covered by two lines
- **Leave the elements covered by just one line unchanged**
- Subtract 1 from the uncovered elements

Giving

	Cut	Sew	Fill	Finish	Pack
Ben	1	0	1	0	2
Ellie	0	0	8	2	0
Greg	1	0	0	2	1
Hyo	3	0	1	4	0
Toby	1	0	0	1	2

We need five lines to cover all the zeros, so we have found our optimal solution.

In fact there are two solutions:

Either

Ben – Finish,   Ellie – Cut,   Greg – Fill,   Hyo – Pack,   Toby – Sew

Or

Ben – Finish,   Ellie – Cut,   Greg – Sew,   Hyo – Pack,   Toby – Fill

Both solutions have a total time of 29 seconds.
(6 + 5 + 3 + 7 + 8, or 6 + 5 + 9 + 7 + 2)

To find the solution from the final table we just use logic – treat it as a logic problem.

First locate any single zeros in rows or columns since these must be used.

	Cut	Sew	Fill	Finish	Pack
Ben	1	0	1	0	2
Ellie	0	0	8	2	0
Greg	1	0	0	2	1
Hyo	3	0	1	4	0
Toby	1	0	0	1	2

So we must include Ellie – Cut and Ben – Finish in our solution.

This leaves Greg, Hyo and Toby and Sew, Fill and Pack to be assigned.

The only person now able to do Pack is Hyo, so we assign her.

	Cut	Sew	Fill	Finish	Pack
Ben	1	0	1	0	2
Ellie	0	0	8	2	0
Greg	1	0	0	2	1
Hyo	3	0	1	4	0
Toby	1	0	0	1	2

We now see that Greg and Toby must cover Sew and Fill, which both are able to do – giving our two solutions.

Note that our 'check' no longer works as easily.

The row reductions (there were no column reductions) are

$3 + 5 + 3 + 2 + 2 = 15$.

The first iteration had 3 as its minimum uncovered element, the second 2 and the third 1, but $15 + 3 + 2 + 1 \neq 27$.

There is a way of checking, but it gets more complicated.

If the problem is $n \times n$ and at each stage the number of lines needed to cover the zeros is m, and the smallest uncovered element is e, the cost of the solution is:

Row and column reductions $+ \sum(n - m)e$.

So in this case the check would be:

$15 + 3(3) + 2(2) + 1(1) = 29$

This check will NOT be required in the examination it is included here for interest only.

Exercise 2A

In questions 1 to 4 the tables show the cost, in pounds, of allocating workers to tasks. Reducing rows first, use the Hungarian algorithm to find an allocation that minimises the cost. You should make your method clear, showing the table at each stage and state your final solution and its cost.

1

	Task X	Task Y	Task Z
Worker A	34	35	31
Worker B	26	31	27
Worker C	30	37	32

2

	Task A	Task B	Task C	Task D
Worker P	34	37	32	32
Worker Q	35	32	34	37
Worker R	42	35	37	36
Worker S	38	34	35	39

3

	Task R	Task S	Task T	Task U
Worker J	20	22	14	24
Worker K	20	19	12	20
Worker L	13	10	18	16
Worker M	22	23	9	28

4

	Task V	Task W	Task X	Task Y	Task Z
Worker D	85	95	97	87	80
Worker E	110	115	95	105	100
Worker F	90	95	86	93	105
Worker G	85	83	84	85	87
Worker H	100	100	105	120	95

5

	100 m	Hurdles	200 m	400 m
Ahmed	14	21	37	64
Ben	13	22	40	68
Chang	12	20	38	70
Davina	13	21	39	74

A junior school has to enter four pupils in an athletics competition comprising four events; 100 m sprint, hurdles, 200 m, 400 m. The rules are that each pupil may only enter one event and the winning team is the one whose total time for the four events is the least. The school holds trials and the table shows the time, in seconds, that each of the team members takes. Reducing rows first, use the Hungarian algorithm to determine who should participate in which event in order to minimise the total time.

6

	Beech	Elm	Eucalyptus	Oak	Olive
A	153	87	62	144	76
B	162	105	87	152	88
C	159	84	75	165	79
D	145	98	63	170	85
E	149	94	70	138	82

The table shows the cost, in pounds, of purchasing trees from five local nurseries. A landscape gardener wishes to support each of these local nurseries for the year and so decides to use each nursery to supply one type of tree. He will use equal numbers of each type of tree throughout the year.

Reducing rows first, use the Hungarian algorithm to determine which type of tree should be supplied by which nursery in order to minimise the total cost.

2.3 You can adapt the algorithm to use a dummy location.

■ If you do not have an *n* × *n* problem you handle it in the same way that you would a transportation problem. You introduce a dummy row or column and put zeros in as the elements.

Example 4

	Task A	Task B	Task C
Mark	12	23	15
Nicky	14	21	17
Nigel	13	22	20
Susie	14	24	13

The table shows the time, in minutes, taken for four workers Mark, Nicky, Nigel and Susie to do each of three tasks A, B and C.

Use the Hungarian algorithm to obtain an allocation that minimises the total time.

	Task A	Task B	Task C	Task D
Mark	12	23	15	0
Nicky	14	21	17	0
Nigel	13	22	20	0
Susie	14	24	13	0

There are more workers than tasks so introduce a dummy column so that the number of rows is equal to the number of columns.

Add in a extra column and fill it with zeros, since it will take no time to do.

Continue as usual.

There are no row subtractions to do (each row contains a zero), so we subtract 12, 21, 13 and 0 from the columns.

	Task A	Task B	Task C	Task D
Mark	0	2	2	0
Nicky	2	0	4	0
Nigel	1	1	7	0
Susie	2	3	0	0

It takes four straight lines to cover the zeros, so we have an optimal solution.

Mark – task A　　　　　Nicky – task B
Nigel – task D (dummy)　　Susie – task C

The total time taken = 12 + 21 + 0 + 13 = 46 minutes.

Exercise 2B

The tables show the cost, in pounds, of allocating workers to tasks.

Reducing rows first, use the Hungarian algorithm to find an allocation that minimises the cost. You should make your method clear, show the table at each stage, and state your final solution and its cost.

1

	Task M	Task N
Worker J	23	26
Worker K	26	30
Worker L	29	28

2

	Task W	Task X	Task Y	Task Z
Worker A	31	43	19	35
Worker B	28	46	10	34
Worker C	24	42	13	33

3

	Task R	Task S	Task T
Worker W	81	45	55
Worker X	67	32	48
Worker Y	87	38	58
Worker Z	73	37	60

4

	Task E	Task F	Task G	Task H
Worker P	24	42	32	31
Worker Q	22	39	30	35
Worker R	13	34	22	25
Worker S	19	41	27	29
Worker T	18	40	31	33

2.4 You can adapt the algorithm to manage incomplete data.

■ If it is not possible to assign a task to a given person, enter a large value into the matrix at the appropriate place. This makes these particular assignments 'unattractive'.

Example 5

	New Dell (Chinese)	Aye Full Tower (French)	Hows a curry (Indian)	Piece a Pasta (Italian)
Denis	–	27	15	40
Hilary	14	21	17	13
Robert	20	–	13	–
Trudy	14	24	10	30

An agency needs to assign four relief chefs Denis, Hilary, Robert and Trudy to four restaurants serving Chinese, French, Indian and Italian food. The travelling expenses, in pounds, that will be paid to each chef are shown in the table. Some of the chefs can not work in some of the restaurants, since they are not familiar with that style of cookery.

Use the Hungarian algorithm, reducing rows first, to obtain an allocation that minimises the total travelling expenses paid.

	New Dell (Chinese)	Aye Full Tower (French)	Hows a curry (Indian)	Piece a Pasta (Italian)
Denis	100	27	15	40
Hilary	14	21	17	13
Robert	20	100	13	100
Trudy	14	24	10	30

Replace each piece of missing data by a large number. This number is usually found by (at least) doubling the largest entry. In this case, enter values of 100 in place of each piece of missing data.

We now continue as usual, applying the Hungarian algorithm to the new matrix.

Reducing rows gives:

	New Dell (Chinese)	Aye Full Tower (French)	Hows a curry (Indian)	Piece a Pasta (Italian)
Denis	85	12	0	25
Hilary	1	8	4	0
Robert	7	87	0	87
Trudy	4	14	0	20

Reducing columns gives:

	New Dell (Chinese)	Aye Full Tower (French)	Hows a curry (Indian)	Piece a Pasta (Italian)
Denis	84	4	0	25
Hilary	0	0	4	0
Robert	6	79	0	87
Trudy	3	6	0	20

The zeros can be covered by two lines like this:

	New Dell (Chinese)	Aye Full Tower (French)	Hows a curry (Indian)	Piece a Pasta (Italian)
Denis	84	4	0	25
Hilary	0	0	4	0
Robert	6	79	0	87
Trudy	3	6	0	20

The minimum uncovered element is 3.

	New Dell (Chinese)	Aye Full Tower (French)	Hows a curry (Indian)	Piece a Pasta (Italian)
Denis	81	1	0	22
Hilary	0	0	7	0
Robert	3	76	0	84
Trudy	0	3	0	17

Using the short cut
- add 3 to the cells covered by two lines
- leave the cells covered by one line unchanged
- subtract 3 from the cells uncovered by a line.

The zeros can now be covered with just three lines, like this

	New Dell (Chinese)	Aye Full Tower (French)	Hows a curry (Indian)	Piece a Pasta (Italian)
Denis	81	1	0	22
Hilary	0	0	7	0
Robert	3	76	0	84
Trudy	0	3	0	17

The smallest uncovered element is 1, giving (after applying the short cut again):

	New Dell (Chinese)	Aye Full Tower (French)	Hows a curry (Indian)	Piece a Pasta (Italian)
Denis	81	0	0	21
Hilary	1	0	8	0
Robert	3	75	0	83
Trudy	0	2	0	16

Four lines are needed to cover the zeros, so we have an optimal solution.

Denis – Aye Full Tower   Hilary – Piece a Pasta
Robert – Hows a curry   Trudy – New Dell

The total cost = 27 + 13 + 13 + 14 = £67.

Applying the (more complicated) check
Row and column reductions = 60.
60 + 2(3) + 1(1) = 67.

Exercise 2C

The tables show the cost, in pounds, of allocating workers to tasks.

The crosses '✗' indicate that that worker cannot be assigned to that task.

Reducing rows first, use the Hungarian algorithm to find an allocation that minimises the cost. You should make your method clear, show the table at each stage and state your final solution and its cost.

1

	Task L	Task M	Task N
Worker P	48	34	✗
Worker Q	✗	37	67
Worker R	53	43	56

2

	Task D	Task E	Task F	Task G
Worker R	38	47	55	53
Worker S	32	✗	47	64
Worker T	✗	53	43	✗
Worker U	41	48	52	47

3

	Task P	Task Q	Task R	Task S
Worker A	46	53	67	75
Worker B	48	✗	61	78
Worker C	42	46	53	62
Worker D	39	50	✗	73

4

	Task R	Task S	Task T	Task U	Task V
Worker J	143	112	149	137	✗
Worker K	149	106	153	115	267
Worker L	137	109	143	121	✗
Worker M	157	✗	✗	134	290
Worker N	126	101	132	111	253

2.5 You can modify the algorithm to deal with a maximum profit allocation.

You have seen how to use the Hungarian algorithm to find a minimum cost allocation.
You can modify the matrix so that you can find a maximum profit allocation.

Example 6

	Task W	Task X	Task Y	Task Z
Cherry	12	23	15	40
Jimmy	14	21	17	20
Mac	13	22	20	30
Plum	14	24	13	10

The numbers represent profits, in pounds, so that the profit of Cherry doing task W is £12, etc.
Find the allocation of tasks to people to maximise the total income.

The Hungarian algorithm finds **minimums** so you have to make
all the numbers in the table negative. The algorithm will find
the minimum solution – i.e. the most negative one, which will, of
course, be the maximum solution.

Start off with

	Task W	Task X	Task Y	Task Z
Cherry	−12	−23	−15	−40
Jimmy	−14	−21	−17	−20
Mac	−13	−22	−20	−30
Plum	−14	−24	−13	−10

Subtract the most negative number from each element (in this case you subtract −40, which is the same as adding 40), so that you lose the negative signs

	Task W	Task X	Task Y	Task Z
Cherry	28	17	25	0
Jimmy	26	19	23	20
Mac	27	18	20	10
Plum	26	16	27	30

A short cut would be to subtract all the numbers in the original matrix from the largest number. In this case you could get this matrix by subtracting each element from 40.

Proceed as before.

The reduced matrix (after reducing both rows and columns) is

	Task W	Task X	Task Y	Task Z
Cherry	21	17	21	0
Jimmy	0	0	0	1
Mac	10	8	6	0
Plum	3	0	7	14

Three lines are needed to cover the zeros

	Task W	Task X	Task Y	Task Z
Cherry	21	17	21	0
Jimmy	0	0	0	1
Mac	10	8	6	0
Plum	3	0	7	14

The minimum uncovered element is 3

So the next matrix is

	Task W	Task X	Task Y	Task Z
Cherry	18	17	18	0
Jimmy	0	3	0	4
Mac	7	8	3	0
Plum	0	0	4	14

Three lines are needed again to cover the zeros.

	Task W	Task X	Task Y	Task Z
Cherry	18	17	18	0
Jimmy	0	3	0	4
Mac	7	8	3	0
Plum	0	0	4	14

The minimum uncovered element is 3 again.

The next matrix is

	Task W	Task X	Task Y	Task Z
Cherry	15	17	15	0
Jimmy	0	6	0	7
Mac	4	8	0	0
Plum	0	0	1	14

Now four lines are needed to cover the zeros as shown.

The optimal solution is

Cherry does Z   Jimmy does W   Mac does Y   Plum does X

At a total profit of   40 + 14 + 20 + 24 = £98

A useful tip

After the row and column reductions, you had this matrix, which needs three lines to cover all the zeros.

	Task W	Task X	Task Y	Task Z
Cherry	21	17	21	0
Jimmy	0	0	0	1
Mac	10	8	6	0
Plum	3	0	7	14

We chose the lines like this:

	Task W	Task X	Task Y	Task Z
Cherry	21	17	21	0
Jimmy	0	0	0	1
Mac	10	8	6	0
Plum	3	0	7	14

This pattern gave the minimum uncovered element as 3, and you needed another iteration to locate the solution.

If you had chosen this pattern of lines

	Task W	Task X	Task Y	Task Z
Cherry	21	17	21	0
Jimmy	0	0	0	1
Mac	10	8	6	0
Plum	3	0	7	14

The minimum uncovered element would have been 6 and we get the following matrix.

	Task W	Task X	Task Y	Task Z
Cherry	15	11	15	0
Jimmy	0	0	0	7
Mac	4	2	0	0
Plum	3	0	7	20

This also gives the optimal solution, but with one fewer iteration.

So if there is a choice in the pattern of lines, without increasing the number of lines, choose the pattern that leaves the minimum uncovered element as large as possible. It is likely to lead to a speedier solution.

Exercise 2D

The tables show the **profit**, in pounds, of allocating workers to tasks.

Reducing rows first, use the Hungarian algorithm to find an allocation that maximises the profit. You should make your method clear, show the table at each stage and state your final solution and its profit.

1

	Task C	Task D	Task E
Worker L	37	15	12
Worker M	25	13	16
Worker N	32	41	35

2

	Task S	Task T	Task U	Task V
Worker C	36	34	32	35
Worker D	37	32	34	33
Worker E	42	35	37	36
Worker F	39	34	35	35

3

	Task E	Task F	Task G	Task H
Worker R	20	22	14	24
Worker S	20	19	12	20
Worker T	13	10	18	16
Worker U	22	23	9	28

4

	Task J	Task K	Task L	Task M	Task N
Worker A	85	95	86	87	97
Worker B	110	111	95	115	100
Worker C	90	95	86	93	105
Worker D	85	87	84	85	87
Worker E	100	100	105	120	95

2.6 You can formulate allocation problems as linear programming problems.

In an allocation problem, you match one person to just one task and each task to just one person.

You have just two options for the decision variables, either the person is going to do the task or they are not.

You use **binary coding** to signal this, with 1 representing allocating that person to the task and 0 representing not allocating them to the task.

■ **There is a standard way of presenting an allocation problem as a linear programming problem:**
   - **first define your decision variables,**
   - **next write down the objective function,**
   - **finally write down the constraints.**

Example 7

Formulate the allocation problem from example 1 as a linear programming problem.

	Task W	Task X	Task Y	Task Z
Kris	12	23	15	40
Laura	14	21	17	20
Sam	13	22	20	30
Steve	14	24	13	10

The decision variables are:

Let x_{ij} be 0 or 1

$$x_{ij} \begin{cases} 1 \text{ if worker } i \text{ does job } j \\ 0 \text{ otherwise} \end{cases}$$

where $i \in \{$Kris, Laura, Sam, Steve$\}$ and $j \in \{$W, X, Y, Z$\}$

The objective is to minimise the total cost, which will be calculated by finding the sum of the product of the cost and the allocation.

In this case the objective is

$$\text{Minimise } C = 12x_{11} + 23x_{12} + 15x_{13} + 40x_{14}$$
$$+ 14x_{21} + 21x_{22} + 17x_{23} + 20x_{24}$$
$$+ 13x_{31} + 22x_{32} + 20x_{33} + 30x_{34}$$
$$+ 14x_{41} + 24x_{42} + 13x_{43} + 10x_{44}$$

The constraints can be found as follows.

Since each person is allocated to just one task each row will contain precisely one 1 and all other entries will be 0.

If

$x_{Kris,Z} = 1$ and $x_{Sam,W} = 0$

this would mean that Kris is allocated to task Z and Sam is not allocated to task W.

The notation is usually simplified and refers to the row number and column number rather than the names.

So $x_{Kris,Z} = 1$ and $x_{Sam,W} = 0$ would be replaced by $x_{14} = 1$ and $x_{31} = 0$

since Kris and Sam are in rows 1 and 3 and Z and W are in columns 4 and 1 respectively.

Consider Kris. You know she will do just one task, so just one variable in her row will be 1, all the rest will be zero.

Hence

$$x_{11} + x_{12} + x_{13} + x_{14} = 1$$

Similarly, since each task will be done by just one person each column will contain precisely one 1 and all other entries will be O.

Consider Y. You know that just one person will be allocated to it, so just one variable in the column will be 1, all the rest will be zero.

Hence

$$x_{13} + x_{23} + x_{33} + x_{43} = 1$$

You can write similar constraints for each person and each task. This enables you to formulate the allocation problem as a linear programming problem.

Example 8

	A	B	C	D
1	10	2	8	6
2	9	3	11	3
3	3	1	4	2
4	3	2	1	5

Four machines 1, 2, 3 and 4 are to be used to perform four tasks, A, B, C and D. Each machine is to be assigned to just one task and each task must be assigned to just one machine. The cost, in thousands of pounds, of using each machine for each task is given in the table above.

Formulate the above assignment problem as a linear programming problem, defining your variables and making the objective and constraints clear.

FIRST define your decision variables:

Let x_{ij} be O or 1

$$x_{ij} \begin{cases} 1 \text{ if machine } i \text{ does task } j \\ 0 \text{ otherwise} \end{cases}$$

where $i \in \{1, 2, 3, 4\}$ and $j \in \{A, B, C, D\}$

NEXT write down the objective function:

$$\begin{aligned} \text{minimise } C = {}& 10x_{1A} + 2x_{1B} + 8x_{1C} + 6x_{1D} \\ &+ 9x_{2A} + 3x_{2B} + 11x_{2C} + 3x_{2D} \\ &+ 3x_{3A} + x_{3B} + 4x_{3C} + 2x_{3D} \\ &+ 3x_{4A} + 2x_{4B} + x_{4C} + 5x_{4D} \end{aligned}$$

FINALLY write down the constraints:

$$\begin{aligned} \text{Subject to } \quad & x_{1A} + x_{1B} + x_{1C} + x_{1D} = 1 && \text{or} \quad \sum x_{1j} = 1 \\ & x_{2A} + x_{2B} + x_{2C} + x_{2D} = 1 && \text{or} \quad \sum x_{2j} = 1 \\ & x_{3A} + x_{3B} + x_{3C} + x_{3D} = 1 && \text{or} \quad \sum x_{3j} = 1 \\ & x_{4A} + x_{4B} + x_{4C} + x_{4D} = 1 && \text{or} \quad \sum x_{4j} = 1 \\ & x_{1A} + x_{2A} + x_{3A} + x_{4A} = 1 && \text{or} \quad \sum x_{iA} = 1 \\ & x_{1B} + x_{2B} + x_{3B} + x_{4B} = 1 && \text{or} \quad \sum x_{iB} = 1 \\ & x_{1C} + x_{2C} + x_{3C} + x_{4C} = 1 && \text{or} \quad \sum x_{iC} = 1 \\ & x_{1D} + x_{2D} + x_{3D} + x_{4D} = 1 && \text{or} \quad \sum x_{iD} = 1 \end{aligned}$$

■ **If you are required to formulate the MAXIMISING problem as a linear programming problem, transform the matrix as on pages 48–9, then proceed as usual.**

Example 9

	1	2	3
A	11	15	21
B	14	18	17
C	16	13	23

Three workers A, B and C are to be assigned to three sites 1, 2 and 3 in order to collect the names and addresses of those who may be considering changing their fuel supplier. The number of names and addresses each worker is likely to collect at each site is shown in the table above.

Each worker must be assigned to just one site and each site assigned to just one worker. The fuel company wish to **maximise** the number of names and addresses collected.

Formulate this as a linear programming problem, defining your variables and making the objective and constraints clear.

Before we formulate the problem, we need to alter the matrix because it is a maximising problem.

We simply subtract all entries from the largest number giving

	1	2	3
A	12	8	2
B	9	5	6
C	7	10	0

We then proceed as usual.

FIRST define your decision variables:

Let x_{ij} be 0 or 1

$$x_{ij} \begin{cases} 1 \text{ if worker } i \text{ does job } j \\ 0 \text{ otherwise} \end{cases}$$

where $i \in A, B, C$ and $j \in 1, 2, 3$

NEXT write down the objective function:

minimise $P = 12x_{1A} + 8x_{1B} + 2x_{1C}$
$+ 9x_{2A} + 5x_{2B} + 6x_{2C}$
$+ 7x_{3A} + 10x_{3B}$

FINALLY write down the constraints:

Subject to $\quad x_{A1} + x_{A2} + x_{A3} = 1 \quad$ or $\quad \sum x_{Aj} = 1$

$x_{B1} + x_{B2} + x_{B3} = 1 \quad$ or $\quad \sum x_{Bj} = 1$

$x_{C1} + x_{C2} + x_{C3} = 1 \quad$ or $\quad \sum x_{Cj} = 1$

$x_{A1} + x_{B1} + x_{C1} = 1 \quad$ or $\quad \sum x_{i1} = 1$

$x_{A2} + x_{B2} + x_{C2} = 1 \quad$ or $\quad \sum x_{i2} = 1$

$x_{A3} + x_{B3} + x_{C3} = 1 \quad$ or $\quad \sum x_{i3} = 1$

Exercise 2E

Questions 1 and 2

The tables show the cost, in pounds, of allocating workers to tasks.

You wish to minimise the total cost.

Formulate these as linear programming problems, defining your variables and making the objective and constraints clear.

1

	Task C	Task D	Task E
Worker L	37	15	12
Worker M	25	13	16
Worker N	32	41	35

2

	Task S	Task T	Task U	Task V
Worker C	36	34	32	35
Worker D	37	32	34	33
Worker E	42	35	37	36
Worker F	39	34	35	35

Questions 3 and 4
Repeat questions 1 and 2, but take the entries to be the **profit** earned in allocating workers to tasks, and seek to maximise the total profit.

Mixed exercise 2F

1

	Airport	Depot	Docks	Station
Bring-it	322	326	326	328
Collect-it	318	325	324	325
Fetch-it	315	319	317	320
Haul-it	323	322	319	321

A museum is staging a special exhibition. They have been loaned exhibits from other museums and from private collectors. Seven days before the exhibition starts these exhibits will be arriving at the airport, road depot, docks and railway station and in each case the single load has to be transported to the museum. There are four local companies that could deliver the exhibits: Bring-it, Collect-it, Fetch-it and Haul-it. Since all four companies are helping to sponsor the exhibition, the museum wishes to use all four companies, allocating each company to just one arrival point.

The table shows the cost, in pounds, of using each company for each task. The museum wishes to minimise its transportation costs.

Reducing rows first, use the Hungarian algorithm to determine the allocation that minimises the total cost. You must make your method clear and show the table after each stage. State your final allocation and its cost.

2

	Back	Breast	Butterfly	Crawl
Jack	18	20	19	14
Kyle	19	21	19	14
Liam	17	20	20	16
Mike	20	21	20	15

A medley relay swimming team consists of four swimmers. The first member of the team swims one length of backstroke, then the second person swims a length of breaststroke, then the next a length of butterfly and finally the fourth person a length of crawl. Each member of the team must swim just one length. All the team members could swim any of the lengths, but some members of the team are faster at one or two particular strokes.

The table shows the time, in seconds, each member of the team took to swim each length using each type of stroke during the last training session.

a Use the Hungarian algorithm, reducing rows first, to find an allocation that minimises the total time it takes the team to complete all four lengths.

b State the best time in which this team could complete the race.

c Show that there is more than one way of allocating the team so that they can achieve this best time.

> In fact there are four optimal solutions to this problem.

3

	Grand Hall	Dining Room	Gallery	Bedroom	Kitchen
Alf	8	19	11	14	12
Betty	12	17	14	18	20
Charlie	10	22	18	14	19
Donna	9	15	16	15	21
Eve	14	23	20	20	19

Five tour guides work at Primkal Mansion. They talk to groups of tourists about five particularly significant rooms. Each tour guide will be stationed in a particular room for the day, but may change rooms the next day. The tourists will listen to each talk before moving on to the next room. Once they have listened to all five talks they will head off to the gift shop.

The table shows the average length of each tour guide's talk in each room.

A tourist party arrives at the Mansion.

a Use the Hungarian algorithm, reducing rows first, to find the **quickest** time that the tour could take. You should state the optimal allocation and its length and show the state of the table at each stage.

b Adapt the table and re-apply the Hungarian algorithm, reducing rows first, to find the **longest** time that the tour could take. You should state the optimal allocation and its duration and show the state of the table at each stage.

4

	Award ceremony	Film premiere	Celebrity party
Denzel	245	378	459
Eun-Ling	250	387	467
Frank	224	350	442
Gabby	231	364	453

A company hires out chauffer-driven, luxury stretch-limousines. They have to provide cars for three events next Saturday night: an award ceremony, a film premiere and a celebrity party. The company has four chauffeurs available and the cost, in pounds, of assigning each of them to each event is shown in the table above. The company wishes to minimise its total costs.

a Explain why it is necessary to add a dummy event.

b Reducing rows first, use the Hungarian algorithm to determine the allocation that minimises the total cost. You should state the optimal allocation and its cost and show the state of the table at each stage.

5

	Catering	Cleaning	Computer	Copying	Post
Blue	No	863	636	628	739
Green	562	796	583	478	674
Orange	No	825	672	583	756
Red	635	881	650	538	No
Yellow	688	934	No	554	No

A large office block is to be serviced and supplied by five companies Blue supplies, Green services, Orange office supplies, Red Co and Yellow Ltd. These companies have each applied to take care of catering, cleaning, computer supplies/servicing, copying and postal services.

The table shows the daily cost of using each firm, in pounds.

For political reasons the owners of the office block will use all five companies, one for each of the five tasks.

Some of the companies cannot offer some services and this is indicated by 'No'.

Use the Hungarian algorithm, reducing rows first, to allocate the companies to the servies in such a way as to minimise the total cost. You should state the optimal allocation and its cost and show the state of the table at each stage.

6

	Cafe	Coffee shop	Restaurant	Snack shop
Ghost train	834	365	580	648
Log flume	874	375	No	593
Roller coaster	743	289	No	665
Teddie's adventure	899	500	794	No

The owners of a theme park wish to provide a café, coffee shop, restaurant and snack shop at four sites: next to the ghost train, log flume, roller coaster and teddie's adventure. They employ a market researcher who estimates the daily profit of each type of catering at each site.

The market researcher also suggests that some types of catering are not suitable at some of the sites, these are indicated by 'No'.

Using the Hungarian algorithm, determine the allocation that provides the maximum daily profit.

This question is a maximising question and one with incomplete data. You need to chose numbers to put at the sites marked 'No' so that they become 'unattractive' to the algorithm after it has been altered to look for the maximum solution.

7

	1	2	3	4
P	143	243	247	475
Q	132	238	218	437
R	126	207	197	408
S	138	222	238	445

Four workers P, Q, R and S are to be assigned to four tasks 1, 2, 3 and 4. Each worker is to be assigned to one task and each task must be assigned to one worker. The cost, in pounds, of using each worker for each task is given in the table above. The cost is to be minimised.

Formulate this as a linear programming problem, defining your variables and making the objective and constraints clear.

8

	A	B	C	D
P	13	17	15	18
Q	15	19	12	19
R	16	20	13	22
S	14	15	17	24

Krunchy Cereals Ltd will send four salesmen P, Q R and S to visit four store managers at A, B, C and D to take orders for their new products. Each salesman will visit only one store manager and each store manager will be visited by just one salesman. The expected value, in thousands of pounds, of the orders won is shown in the table above. The company wishes to maximise the value of the orders.

Formulate this as a linear programming problem, defining your variables and making the objective and constraints clear.

Summary of key points

1 To reduce a cost matrix:
- Subtract the least value in each row from each element of that row.
- Using the new matrix, subtract the least value in each column from each element in that column.

2 The Hungarian algorithm:
- Find the reduced cost matrix.
- Find the minimum number of straight lines (horizontal or vertical) which will cover all of the zeros in the matrix.
- In an $n \times n$ matrix, if you cannot cover the zeros in fewer than n lines, you have an optimal solution and you stop.
- In an $n \times n$ matrix, if you can cover the zeros in fewer than n lines, the solution can be improved.
- Draw in the lines and look for the smallest uncovered element, e.
- Add e to the elements in each covered row and each covered column, adding it twice to any element covered twice.
- Subtract e from every element in the matrix.
- Repeat procedure until an optimal solution is found.

3 As a short cut, after drawing in the lines:
- Increase each element covered by two lines by e.
- Leave each element covered by one line unchanged.
- Subtract e from each uncovered element.

4 If the problem is not $n \times n$ you add a dummy row or column of zeros and proceed as for an $n \times n$ matrix.

5 If the data is incomplete, add a large value in the matrix at an appropriate place. Use (at least) double the largest entry.

6 To find **maximums** begin by making all the numbers in the table **negative**. Subtract the most negative number from each element then proceed as above.

7 To formulate allocation problems as linear programming problems, use binary coding, with 1 representing allocating the person to a task and 0 representing not allocating the person to a task.

8 Present an allocation problem as a linear programming problem by:
- defining decision variables
- writing down the objective function
- writing down the constraints.

In this chapter you will:

- learn the differences between the classical and practical problems
- use a minimum spanning tree method to find an upper bound and a lower bound
- use the nearest neighbour algorithm to find an upper bound.

The travelling salesman problem

This is similar to the route inspection problem in book D1, but here you seek to visit every **node** rather than every **arc**.

You wish to minimise the total distance travelled by a 'salesman' who leaves his home, visits several places and then returns home.

3.1 You understand the terminology used.

You are seeking a **walk** that gives us a minimum **tour**.

- A **walk** in a network is a finite sequence of edges such that the end vertex of one edge is the start vertex of the next.

- A walk which visits every vertex, returning to its starting vertex, is called a **tour.**

There is no viable algorithm known for the solution of the travelling salesman problem. You therefore make use of an heuristic algorithm (this means that it is an intuitive algorithm) which should give a good answer but probably not the optimal answer.

In practice we can find an **upper bound** and **lower bound** for the solution and use these to 'trap' the optimal solution. If our upper and lower bounds are close, then a solution between the two may be acceptable.

You will therefore find upper bounds and then select the **smallest** and find lower bounds and select the **largest**, trying to 'trap' the optimal solution in as narrow an interval as possible.

> If you know that your shortest route is between 123 and 145 miles, say, and you find a route that is 123 miles long, then you know you have found the optimal route. If you find a route that is 130 miles long you may decide that it is 'optimal enough' and use it.

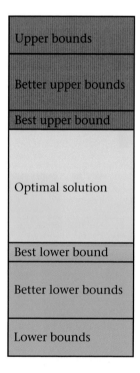

Upper bounds

Better upper bounds

Best upper bound

Optimal solution

Best lower bound

Better lower bounds

Lower bounds

3.2 You should know the differences between the classical and practical problems.

There are two travelling salesman problems:

the **classical** problem in which once a vertex has been visited it may not be revisited, and the **practical** problem in which a vertex may be revisited.

The difference is that:

- ■ in the **classical** problem you must visit each vertex **only once**, before returning to the start, in the **practical** problem you must visit each vertex **at least once** before returning to the start.

3.3 You can convert a network into a complete network of least distances.

If you convert the network into a complete network of least distances, the classical and practical travelling salesman problems are the same.

To create a complete network of least distances you ensure that the **triangle inequality** holds for all triangles in the network.

■   **The triangle inequality states**

> **the longest side of any triangle ⩽ the sum of the two shorter sides.**

(If you had three rods with lengths 4 m, 2 m and 1 m, you could not form a triangle with them. The two shorter rods could not connect up.)

> Equality is possible. There is active debate about whether the resulting figure is a triangle. For the purposes of this topic, equality **would** be acceptable.

If you have a network where the triangle inequality does not hold in one or more triangles, you simply replace the longest arc in those triangles by the sum of the two smaller ones, thereby creating a network which shows the shortest distances.

> In the examination this can usually be done by observation and it is not necessary to apply Dijkstra's algorithm. If it is necessary to use Dijkstra's algorithm you would be directed to do so.

Example 1

Create a table of least distances for the network below.

> Usually in the examination you are asked to **complete** a table of least distances, by finding just a few missing values.

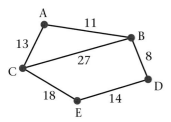

> We need to look carefully for the shortest route between each pair of vertices. **This may not always be the direct route**, so you need to do some careful arithmetic.

> In the examination you are likely to be asked for the 'tricky' ones – for example BC in this first example. So do check carefully for non-direct, shorter routes.

The top row shows the shortest routes starting at A.

	A	B	C	D	E
A	–	11	13	19	31
B		–			
C			–		
D				–	
E					–

Look at routes from A.
There are direct routes from AB and AC and you can see these are the shortest routes.
Using inspection you complete the table to show the least routes AD and AE.
You need to check AE carefully. Using ACE the route is 31. Using ABDE the route is 33, so you record 31 as the value.
This enables you to complete the top row of the table.

You can use this to fill in the first column

	A	B	C	D	E
A	–	11	13	19	31
B	11	–			
C	13		–		
D	19			–	
E	31				–

Since this is not a directed network, the shortest distance from A to D is the same as the shortest distance from D to A. In this topic directed networks will not be examined in D2.

Move on to the remaining routes starting from B.

	A	B	C	D	E
A	–	11	13	19	31
B	11	–	24	8	22
C	13	24	–		
D	19	8		–	
E	31	22			–

The direct route BC on the network is given as 27, but if you use BAC as your route you get 24, so 24 is the least distance from B to C.

You complete BD and BE by observation, using BDE as the shortest route from B to E.

Once again you use the table's symmetry to complete the second column.

Considering the remaining routes starting at C, you get

	A	B	C	D	E
A	–	11	13	19	31
B	11	–	24	8	22
C	13	24	–	32	18
D	19	8	32	–	
E	31	22	18		–

Starting at C you have two routes to find: CD and CE. The direct arc CE is the shortest route.
For CD you need to check CED (length 32) and CABD (length 32), so you can record 32. (You do not need to check CBD since you found that the direct route CB was longer than the route CAB earlier.)

Finally you complete the last entries, giving the completed table of least differences.

	A	B	C	D	E
A	–	11	13	19	31
B	11	–	24	8	22
C	13	24	–	32	18
D	19	8	32	–	14
E	31	22	18	14	–

In the examination candidates will just have to show the final table.

Example 2

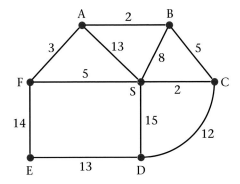

The network above shows the distances, in km, between the central sorting office at S and six post offices A, B, C, D, E and F.

	S	A	B	C	D	E	F
S	–			2			5
A		–	2		19	17	3
B		2	–	5	17		
C	2		5	–	12	21	7
D		19	17	12	–	13	
E		17	19	21	13	–	14
F	5	3		7		14	–

The table shows a partially completed table of least distances.

Complete the table of least distances for the network above, stating your shortest route for each of the entries.

SA – the shortest route is SFA length 8
SB – the shortest route is SCB length 7
SD – the shortest route is SCD length 14
SE – the shortest route is SFE length 19
AC – the shortest route is ABC length 7
BE – the shortest route is BAFE length 19
BF – the shortest route is BAF length 5
DF – the shortest route is DCSF length 19

Which gives the following table of least distances.

	S	A	B	C	D	E	F
S	–	8	7	2	14	19	5
A	8	–	2	7	19	17	3
B	7	2	–	5	17	19	5
C	2	7	5	–	12	21	7
D	14	19	17	12	–	13	19
E	19	17	19	21	13	–	14
F	5	3	5	7	19	14	–

> As the table is completed, the earlier results, as well as the given results, can be used to help with later ones.

Exercise 3A

Photocopy masters are available on the CD-ROM for each of these questions.

Questions 1 to 4.

Complete the table of least distances for each of the networks. State the route you used for each of your entries.

1

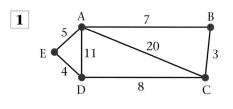

Table of least differences

	A	B	C	D	E
A	–	7			5
B	7	–	3		
C		3	–	8	12
D			8	–	4
E	5		12	4	–

2

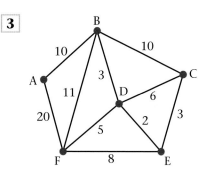

Table of least differences

	A	B	C	D	E
A	–	5		7	
B	5	–		2	
C			–	5	
D	7	2	5	–	2
E				2	–

3

Table of least differences

	A	B	C	D	E	F
A	–	10		13	15	
B	10	–		3		
C			–		3	
D	13	3		–	2	5
E	15		3	2	–	
F				5		–

4

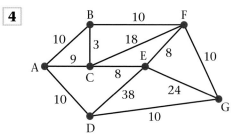

Table of least differences

	A	B	C	D	E	F	G
A	–	10	9	10	17		
B	10	–	3	20		10	20
C	9	3	–	19	8		
D	10	20	19	–		20	10
E	17		8		–	8	18
F		10		20	8	–	10
G		20		10	18	10	–

3.4 You can use a minimum spanning tree method to find an upper bound.

The minimum spanning tree method to find an upper bound

- Find the minimum spanning tree for the network (using Prim's algorithm or Kruskal's algorithm). This guarantees that each vertex is included.

> This method finds an upper bound for the practical problem.

- Double this minimum connector (in effect you keep on retracing your steps) so that completing the cycle is guaranteed.

> You are seeking a **minimum** route. There is therefore a logic in trying to use, as a starting point, the **minimum** spanning tree, which you know how to find.

- Finally seek 'short cuts'. (Make use of some of the non-included arcs that enable you to bypass a repeat of some of the minimum spanning tree.)

Finding an initial upper bound

If you find a minimum spanning tree for a network, an initial attempt to locate an upper bound can be made by finding the minimum spanning tree and repeating each arc.

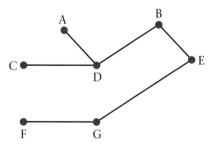

Minium spanning tree   Initial upper bound

This visits each vertex and returns to the start.

You find the length of this route, by doubling the weight of the minimum spanning tree, and using it as an initial upper bound.

■ **An initial upper bound is found by finding the weight of the minimum spanning tree for the network and doubling it.**

Example **3**

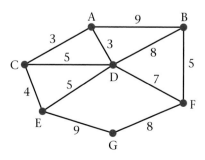

Use Kruskal's algorithm to find a minimum spanning tree for the network above. Hence find an initial upper bound for the travelling salesman problem.

In the examination you could be asked to use Prim's algorithm or Kruskal's algorithm to find the minimum spanning tree. If you are not directed to a particular algorithm, you may use either.

Using Kruskal's algorithm
Putting the arcs in order: {AC, AD}, CE, {BF, CD, DE} DF, {BD, FG}, {AB, EG}
Include AC, AD, CE, BF, reject CD, reject DE, include DF, reject BD, include FG. Tree complete

This gives the following minimum spanning tree

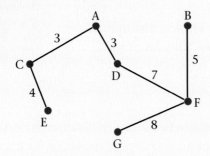

The weight of the minimum spanning tree = 3 + 3 + 4 + 5 + 7 + 8 = 30
The initial upper bound is 2 × 30 = 60.

Example 4

	A	B	C	D	E
A	–	11	13	19	31
B	11	–	24	8	22
C	13	24	–	32	18
D	19	8	32	–	14
E	31	22	18	14	–

Use the table of least distances to find an initial upper bound for the travelling salesman problem in example 1.

Starting at A.

	1	2	3	4	5
	A	B	C	D	E
A	–	11	13	19	31
B	⑪	–	24	8	22
C	⑬	24	–	32	18
D	19	⑧	32	–	14
E	31	22	18	⑭	–

Order of arc inclusion: AB, BD, AC, DE

When using a table it is easier to use Prim's algorithm.

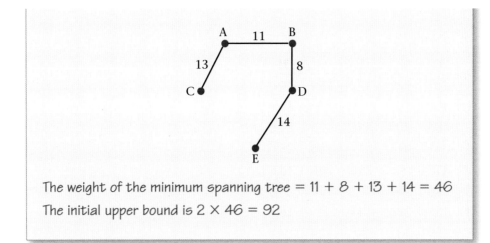

The weight of the minimum spanning tree = 11 + 8 + 13 + 14 = 46
The initial upper bound is 2 × 46 = 92

The initial upper bound is not very good, since we repeat each arc.

■ **We can improve the initial upper bound by looking for short cuts.**

Example 5

Starting from the initial upper bound found in example 4, use a short cut to reduce the upper bound to below 70.

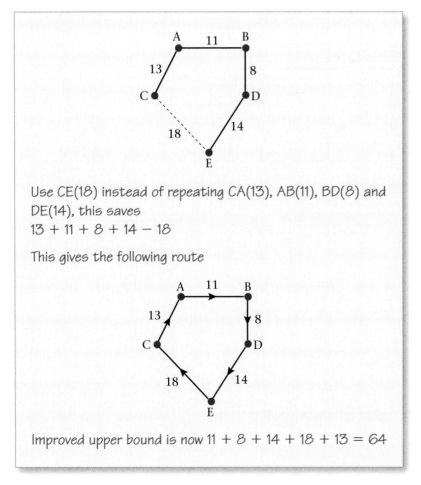

Use CE(18) instead of repeating CA(13), AB(11), BD(8) and DE(14), this saves
13 + 11 + 8 + 14 − 18

This gives the following route

Improved upper bound is now 11 + 8 + 14 + 18 + 13 = 64

Example 6

	A	B	C	D	E	F	G	H
A	–	47	84	382	120	172	299	144
B	47	–	121	402	155	193	319	165
C	84	121	–	456	200	246	373	218
D	382	402	456	–	413	220	155	289
E	120	155	200	413	–	204	286	131
F	172	193	246	220	204	–	144	70
G	299	319	373	155	286	144	–	160
H	144	165	218	289	131	70	160	–

The table shows the distance, in miles, between eight cities. A politician has to visit each city, starting and finishing at A. She wishes to minimise the total distance travelled.

a Find a minimum spanning tree for this network.

b Hence find an upper bound for this problem.

c Use short cuts to reduce this upper bound to a value below 1300.

 (adapted)

a Using Prim's algorithm

	1	2	3	8	4	6	7	5
	A	B	C	D	E	F	G	H
A	–	47	84	382	120	172	299	144
B	(47)	–	121	402	155	193	319	165
C	(84)	121	–	456	200	246	373	218
D	382	402	456	–	413	220	(155)	289
E	(120)	155	200	413	–	204	286	131
F	172	193	246	220	204	–	144	(70)
G	299	319	373	155	286	(144)	–	160
H	144	165	218	289	(131)	70	160	–

Order of arc selection: AB, AC, AE, EH, HF, FG, GD

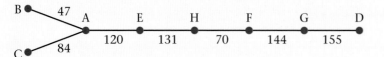

b The initial upper bound is 2 × 751 = 1502

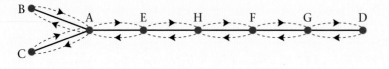

c Looking at the tree, likely shortcuts are AD and BC.

AD saves 120 + 131 + 70 + 144 + 155 − 382 = 238

BC saves 47 + 84 − 121 = 10

This leaves the following tour

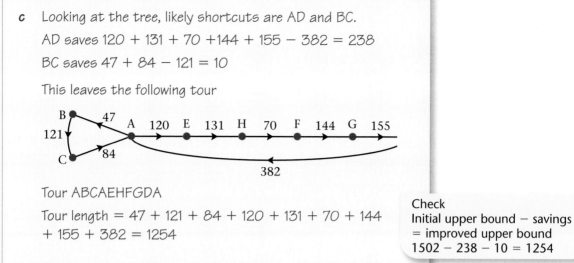

Tour ABCAEHFGDA

Tour length = 47 + 121 + 84 + 120 + 131 + 70 + 144
+ 155 + 382 = 1254

> **Check**
> Initial upper bound − savings
> = improved upper bound
> 1502 − 238 − 10 = 1254

Selecting the better upper bound

In example 6 there are many other shortcuts that could be tried. For example, here are three others.

> The shortcut from A to D alone would have produced an improved upper bound of 1264.

i BD saves 265, so this shortcut alone would have been sufficient

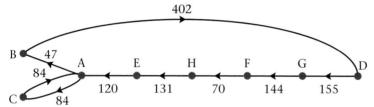

Tour ABDGFHEACA, length 1237

ii CB + AF + FD

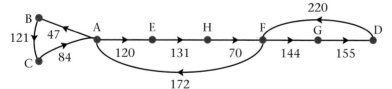

Tour ABCAEHFGDFA, length 1264

iii CD alone saves 248

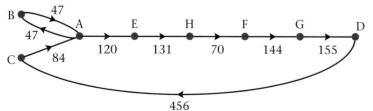

Tour ABAEHFGDCA, length 1254

Of these the best upper bound is 1237 since this is the smallest.

■ **Aim to make the upper bound as low as possible to reduce the interval in which the optimal solution is contained.**

Exercise 3B

1

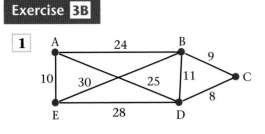

a Find a minimum spanning tree for the network above and hence find an initial upper bound for the travelling salesman problem.

b Use a shortcut to find a better upper bound.

c State the route given by your improved upper bound and state its length.

2

	A	B	C	D	E
A	–	13	11	19	14
B	13	–	12	7	16
C	11	12	–	11	8
D	19	7	11	–	14
E	14	16	8	14	–

A council employee needs to service five sets of traffic lights located at A, B, C, D and E. The table shows the distance, in miles between the lights. She will start and finish at A and wishes to minimise her total travelling distance.

a Find the minimum spanning tree for the network.

b Hence find an initial upper bound for the length of the employee's route.

c Use shortcuts to reduce the upper bound to a value below 65.

d State the route given by your improved upper bound and state its length.

3

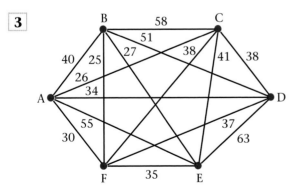

a Find a minimum spanning tree for the network above and hence find an initial upper bound for the travelling salesman problem.

b Use shortcuts to reduce the upper bound to below 240.

c State the route given by your improved upper bound and state its length.

4		S	V	W	X	Y	Z
	S	–	75	30	55	70	70
	V	75	–	55	30	40	15
	W	30	55	–	65	45	55
	X	55	30	65	–	15	10
	Y	70	40	45	15	–	20
	Z	70	15	55	10	20	–

The table shows the time, in minutes, taken to travel between a surgery S and five farms V, W, X, Y and Z. A vet needs to visit animals at each of the farms and wishes to minimise the total travel time. He will start and finish at the surgery, S.

a Find a minimum spanning tree for the network above and hence find an initial upper bound for the travelling salesman problem.

b Use the method of shortcuts to reduce the upper bound to below 200.

c State the route given by your improved upper bound and state its length.

3.5 You can use a minimum spanning tree method to find a lower bound.

This method finds a lower bound for the classical problem.

The ideal solution would be one where each vertex is visited only once, and thus each vertex will have only two 'shortest' arcs incident on it in the solution. If you find the ideal optimal solution it would be possible to remove any vertex and the two arcs linking it into the tour, leaving you with a minimum spanning tree for the remaining vertices. This is the essence of the algorithm.

The minimum spanning tree method to find a lower bound
- Remove each vertex in turn, together with its arcs.
- Find the residual minimum spanning tree (RMST) and its length.
- Add to the RMST the 'cost' of reconnecting the deleted vertex **by the two shortest, distinct, arcs** and note the totals.
- The greatest of these totals is used for the lower bound.
- Make the lower bound as high as possible to reduce the interval in which the optimal solution is contained.
- You have found an optimal solution if
    the lower bound gives a tour, or
    the lower bound has the same value as the upper bound.

This method of deletion and reconnection will not, in general, give a viable tour. The one most likely to be a tour is the greatest, which is why it is selected.

Example 7

	A	B	C	D	E
A	–	11	13	19	31
B	11	–	24	8	22
C	13	24	–	32	18
D	19	8	32	–	14
E	31	22	18	14	–

a By deleting vertex A, find a lower bound to the travelling salesman problem for the network in example 4.

b Comment on your answer.

In the examination you will be directed to delete specific vertices. You will not have to check all the vertices, unless you are specifically instructed to do so.

a When A is deleted, the table for the residual network becomes

	1	4	2	3
	B	C	D	E
B	–	24	8	22
C	24	–	32	(18)
D	(8)	32	–	14
E	22	18	(14)	–

Using Prim's algorithm starting at B, the order of arc selection is
BD, DE and EC
The residual minimum spanning tree is

B ——8—— D ——14—— E ——18—— C

Weight of residual minimum spanning tree = 8 + 14 + 18 = 40
The two least arcs from A are AB (11) and AC (13)

B ——8—— D ——14—— E ——18—— C
  11 ⋯   ⋯ 13
         A

Lower bound = weight of RMST + weights of 2 least arcs from A

= 40 + 11 + 13

= 64

b Either: The lower bound is a tour, therefore it is optimal.

or: The improved upper bound (found in example 5) is 64. Since the improved upper bound = 64 = lower bound, we have found the optimal solution for this travelling salesman problem.

Example 8

	A	B	C	D	E	F	G	H
A	–	47	84	382	120	172	299	144
B	47	–	121	402	155	193	319	165
C	84	121	–	456	200	246	373	218
D	382	402	456	–	413	220	155	289
E	120	155	200	413	–	204	286	131
F	172	193	246	220	204	–	144	70
G	299	319	373	155	286	144	–	160
H	144	165	218	289	131	70	160	–

a By deleting vertices A then G, find two lower bounds to the travelling salesman problem for the network above (from example 6).

b Select the better lower bound of the two found in part **a**, give a reason for your answer.

c Taking your answer to **b** and using the better upper bound, 1237, found on page 71, write down the smallest interval that must contain the length of the optimal route.

> As a class exercise, all vertices could be deleted, in turn, to find other lower bounds.

a i Deleting A and using Prim's algorithm starting at B

	1	**2**	**7**	**3**	**5**	**6**	**4**
	B	**C**	**D**	**E**	**F**	**G**	**H**
B	–	121	402	155	193	319	165
C	(121)	–	456	200	246	373	218
D	402	456	–	413	220	(155)	289
E	(155)	200	413	–	204	286	131
F	193	246	220	204	–	144	(70)
G	319	373	155	286	(144)	–	160
H	165	218	289	(131)	70	160	–

Order of arc selection: BC, BE, EH, HF, FG, GD

> In the examination you need to make your method for finding the RMST clear. The order of arc selection is sufficient to demonstrate that you have applied Prim's algorithm correctly.

The residual minimum spanning tree is:

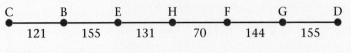

C —121— B —155— E —131— H —70— F —144— G —155— D

Weight of RMST = 776

Two least arcs from A are AB (47) and AC (84)

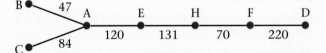

C 121 B 155 E 131 H 70 F 144 G 155 D

47 ⋯ 84

A

Lower bound by deleting A = 776 + 47 + 84 = 907

> This is *not* a tour. In general you do not get a tour when finding a lower bound.

ii Deleting G and using Prim's algorithm starting at A

	1	2	3	7	4	6	5
	A	B	C	D	E	F	H
A	–	47	84	382	120	172	144
B	(47)	–	121	402	155	193	165
C	(84)	121	–	456	200	245	218
D	382	402	456	–	413	(220)	289
E	(120)	155	200	413	–	204	131
F	172	193	246	220	204	–	(70)
H	144	165	218	289	(131)	70	–

Order of arc selection: AB, AC, AE, EH, HF and FD

The residual minimum spanning tree is:

B• 47 A E H F D

C• 84   120 131 70 220

Weight of RMST = 672

Two least arcs are GF (144) and GD (155)

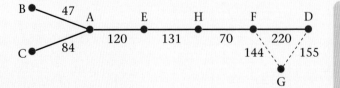

B• 47 A E H F D

C• 84   120 131 70 220

144 ⋯ 155

G

> Once again we do not get a tour when we 'reconnect' G. The lower bound rarely gives a tour. **If it does give a tour we have an optimal solution.**

Lower bound by deleting G = 672 + 144 + 155 = 971

b The better lower bound is the higher one, 971, the one obtained by deleting G.
This will reduce the size of the interval containing the optimal solution.

c The better lower bound is 971, the better upper bound is 1237

$$971 < \text{optimal solution} \leqslant 1237$$

> We use < for the lower bound, because we can see it is not a solution.
> We use ⩽ with the upper bound, because it may be the optimal solution.

Exercise 3C

1

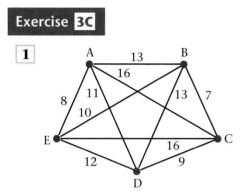

a By deleting vertex A, find a lower bound to the travelling salesman problem for the network above.

b Comment on your answer.

2

	A	B	C	D	E
A	–	13	11	19	14
B	13	–	12	7	16
C	11	12	–	11	8
D	19	7	11	–	14
E	14	16	8	14	–

A council employee needs to service five sets of traffic lights located at A, B, C, D and E. The table shows the distance, in miles between the lights. She will start and finish at A and wishes to minimise her total travelling distance.

a By deleting vertices A then B find two lower bounds for the employee's route.

b Select the better lower bound, giving a reason for your answer.

3

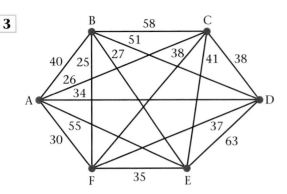

a By deleting vertices A then B, find two lower bounds for the travelling salesman problem.

b Select the better lower bound, giving a reason for your answer.

c Use inequalities, your answer to **b** and the improved upper bound found in exercise 3B Question 3, to write down the smallest interval containing the optimal route

4

	S	V	W	X	Y	Z
S	–	75	30	55	70	70
V	75	–	55	30	40	15
W	30	55	–	65	45	55
X	55	30	65	–	15	10
Y	70	40	45	15	–	20
Z	70	15	55	10	20	–

The table shows the time, in minutes, taken to travel between a surgery S and five farms V, W, X, Y and Z. A vet needs to visit animals at each of the farms and wishes to minimise the total travel time. He will start and finish at the surgery, S.

a By deleting vertices S then V, find two lower bounds for the vet's route.

b Select the better lower bound, giving a reason for your answer.

c Use inequalities, your answer to **b** and the improved upper bound found in exercise 3B Question 4, to write down the smallest interval containing the optimal route.

3.6 You can use the nearest neighbour algorithm to find an upper bound.

The particular value of the upper bound is that it gives a solution, probably not the optimal solution, but a tour that could be used.

It is difficult to find a good set of shortcuts in a large network, so the minimum spanning tree method can be difficult to use.

There are other heuristic algorithms to help find an upper bound and one is the **nearest neighbour algorithm**. The method is as follows.

The nearest neighbour algorithm

1 Select each vertex in turn as a starting point.

2 Go to the nearest vertex which has not yet been visited.

3 Repeat step 2 until all vertices have been visited and then return to the start vertex using the shortest route.

4 Once all vertices have been used as the starting vertex, select the tour with the smallest length as the upper bound.

> In the examination you will be directed to start with specific vertices. You will only have to check all the vertices, if you are specifically instructed to do so.

> **Warning – do not confuse the nearest neighbour and Prim's algorithm**
>
> The nearest neighbour algorithm is what is often done by students who implement Prim's algorithm incorrectly.
>
> In Prim's algorithm you look for the nearest vertex to **any** of the vertices in your growing tree.
>
> In the nearest neighbour you look for the nearest vertex to the **last vertex chosen**.

Example 9

	A	B	C	D	E
A	–	8	7	29	13
B	8	–	9	24	14
C	7	9	–	23	6
D	29	24	23	–	21
E	13	14	6	21	–

Apply the nearest neighbour algorithm, using A then B then C as starting vertices, to find an upper bound to the travelling salesman problem.

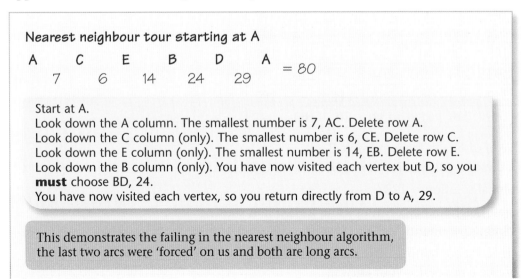

Nearest neighbour tour starting at A

A   C   E   B   D   A   = 80
   7   6   14   24   29

Start at A.
Look down the A column. The smallest number is 7, AC. Delete row A.
Look down the C column (only). The smallest number is 6, CE. Delete row C.
Look down the E column (only). The smallest number is 14, EB. Delete row E.
Look down the B column (only). You have now visited each vertex but D, so you **must** choose BD, 24.
You have now visited each vertex, so you return directly from D to A, 29.

> This demonstrates the failing in the nearest neighbour algorithm, the last two arcs were 'forced' on us and both are long arcs.

Nearest neighbour tour starting at B

B   A   C   E   D   B
  8   7   6   21   24   = 66

We start at B.
Look down the B column. The smallest number is 8, BA. Delete row B.
Look down the A column (only). The smallest number is 7, AC. Delete row A.
Look down the C column (only). The smallest number is 6, CE. Delete row C.
Look down the E column (only). You have now visited each vertex but D, so you
must choose ED, 21.
You have now visited each vertex, so you return directly from D to B, 24.

Nearest neighbour tour starting at C

C   E   A   B   D   C
  6   13   8   24   23   = 74

Start at C.
Look down the C column. The smallest number is 6, CE. Delete row C.
Look down the E column (only). The smallest number is 13, EA. Delete row E.
Look down the A column (only). The smallest number is 8, AB. Delete row A.
Look down the B column (only). You have now visited each vertex but D, so you
must choose BD, 24.
You have now visited each vertex, so you return directly from D to C, 23.

You now have three answers; 80, 66 and 74. Select 66 as the best upper
bound, since this gives the shortest tour.

Example 10

	A	B	C	D	E	F	G	H
A	–	47	84	382	120	172	299	144
B	47	–	121	402	155	193	319	165
C	84	121	–	456	200	246	373	218
D	382	402	456	–	413	220	155	289
E	120	155	200	413	–	204	286	131
F	172	193	246	220	204	–	144	70
G	299	319	373	155	286	144	–	160
H	144	165	218	289	131	70	160	–

This is the table of distances from example 8.
We found that the optimal solution lay in the following interval

$$971 < \text{optimal solution} \leqslant 1237.$$

a Use the nearest neighbour algorithm, using A then B as starting
vertices, to find upper bounds for the travelling salesman problem.

b Review the interval containing the optimal solution and amend it if
necessary, giving a reason for your answer.

As a class exercise,
all vertices could
be used, in turn, to
find other upper
bounds.

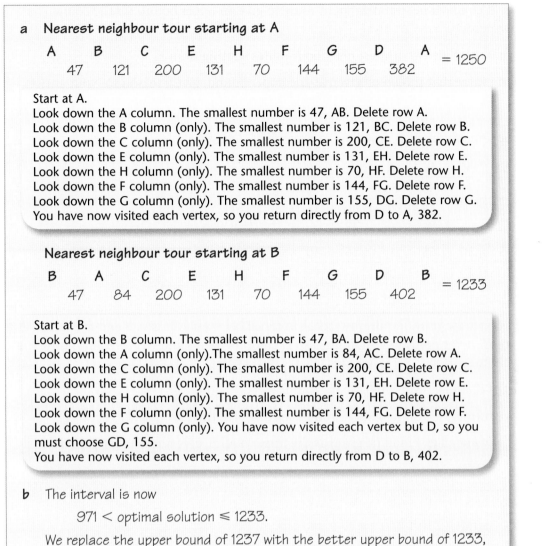

a Nearest neighbour tour starting at A

A		B		C		E		H		F		G		D		A	
	47		121		200		131		70		144		155		382		= 1250

Start at A.
Look down the A column. The smallest number is 47, AB. Delete row A.
Look down the B column (only). The smallest number is 121, BC. Delete row B.
Look down the C column (only). The smallest number is 200, CE. Delete row C.
Look down the E column (only). The smallest number is 131, EH. Delete row E.
Look down the H column (only). The smallest number is 70, HF. Delete row H.
Look down the F column (only). The smallest number is 144, FG. Delete row F.
Look down the G column (only). The smallest number is 155, DG. Delete row G.
You have now visited each vertex, so you return directly from D to A, 382.

Nearest neighbour tour starting at B

B		A		C		E		H		F		G		D		B	
	47		84		200		131		70		144		155		402		= 1233

Start at B.
Look down the B column. The smallest number is 47, BA. Delete row B.
Look down the A column (only). The smallest number is 84, AC. Delete row A.
Look down the C column (only). The smallest number is 200, CE. Delete row C.
Look down the E column (only). The smallest number is 131, EH. Delete row E.
Look down the H column (only). The smallest number is 70, HF. Delete row H.
Look down the F column (only). The smallest number is 144, FG. Delete row F.
Look down the G column (only). You have now visited each vertex but D, so you must choose GD, 155.
You have now visited each vertex, so you return directly from D to B, 402.

b The interval is now

$$971 < \text{optimal solution} \leqslant 1233.$$

We replace the upper bound of 1237 with the better upper bound of 1233, since it is lower. This reduces the interval containing the optimal solution.

Exercise 3D

1 (This is the same problem as described in exercise 3C Question 2)

	A	B	C	D	E
A	–	13	11	19	14
B	13	–	12	7	16
C	11	12	–	11	8
D	19	7	11	–	14
E	14	16	8	14	–

A council employee needs to service five sets of traffic lights located at A, B, C, D and E. The table shows the distance, in miles between the lights. She wishes to minimise her total travelling distance.

a Starting at D, find a nearest neighbour route to give an upper bound for the council employee's route.

b Show that there are two nearest neighbour routes starting from E.

c Select the value that should be given as the upper bound, give a reason for your answer.

2 (This is the same problem as described in exercise 3C Question 4)

	S	V	W	X	Y	Z
S	–	75	30	55	70	70
V	75	–	55	30	40	15
W	30	55	–	65	45	55
X	55	30	65	–	15	10
Y	70	40	45	15	–	20
Z	70	15	55	10	20	–

The table shows the time, in minutes, taken to travel between a surgery S and five farms V, W, X, Y and Z. A vet needs to visit animals at each of the farms and wishes to minimise the total travel time.

a Starting at Z, find a nearest neighbour route.

b Find two further nearest neighbour routes starting at X then V.

c Select the value that should be given as the upper bound, give a reason for your answer.

3

	R	S	T	U	V	W
R	–	150	210	150	120	240
S	150	–	210	120	210	240
T	210	210	–	120	150	180
U	150	120	120	–	180	270
V	120	210	150	180	–	300
W	240	240	180	270	300	–

A printing company prints six magazines R, S, T, U, V and W, each week. The printing equipment needs to be set up differently for each magazine and the table shows the time, in minutes, needed to set up the equipment from one magazine to another. The printer must print magazine R at the start of the first day each week so the equipment is already set up to print magazine R, and must be left set up for magazine R at the end of the week. The other magazines can be printed in any order.

a If the magazines were printed in the order RSTUVWR, how long would it take in total to set up the equipment?

b Show that there are two nearest neighbour routes starting from U.

c Show that there are three nearest neighbour routes starting from V.

d Select the value that should be given as the upper bound, give a reason for your answer.

Mixed exercise 3E

1

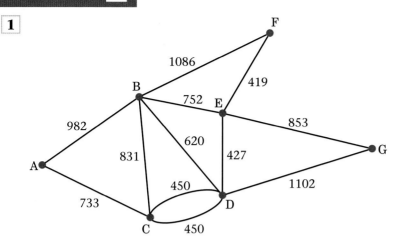

a Use an efficient algorithm to find a minimum connector for the network above. You must make your method clear.

b Hence find an initial upper bound for the travelling salesman problem.

c Use the method of short cuts to find an upper bound below 6100.

2

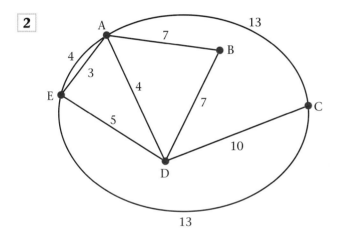

The network above shows a number of hostels in a national park and the possible paths joining them. The numbers on the edges give the lengths, in km, of the paths.

a Draw a complete network showing the shortest distances between the hostels. (You may do this by inspection. The application of an algorithm is not required.)

b Use the nearest neighbour algorithm on the complete network to obtain an upper bound to the length of a tour in this network which starts and finishes at A and visits each hostel exactly once.

c Interpret your result in part **b** in terms of the original network.

3 (This is the network given in example 2)

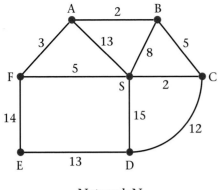

Network N

The table of least distances below was formed from the network, N, above.

	S	A	B	C	D	E	F
S	–	8	7	2	14	19	5
A	8	–	2	7	19	17	3
B	7	2	–	5	17	19	5
C	2	7	5	–	12	21	7
D	14	19	17	12	–	13	19
E	19	17	19	21	13	–	14
F	5	3	5	7	19	14	–

The table shows the distances, in km, between the central sorting office at S and six post offices A, B, C, D, E and F.

A postal worker will leave the sorting office, go to each post office to collect mail and return to the sorting office. He wishes to minimise his route.

a Use Prim's algorithm, starting at S, to obtain two minimum spanning trees. State the order in which you select the arcs.

b Hence find an initial upper bound for the postal worker's route.

c Starting from this upper bound, use shortcuts to reduce the upper bound to a value below 60 km. You must state the shortcuts you use.

d Starting at C, and then at D, find two nearest neighbour routes stating their lengths.

e Select the better upper bound from your answers to **c** and **d**, give a reason for your answer.

f Interpret your answer to **e** in terms of the original network, N, of roads.

g Using the table of least distances, and by deleting C, find a lower bound for the postal worker's route.

4 **a** Explain the difference between the classical and practical travelling salesman problem.

	P	Q	R	S	T	U	V
P	–	19	30	45	38	33	29
Q	19	–	28	27	50	23	55
R	30	28	–	51	29	49	50
S	45	27	51	–	77	21	71
T	38	50	29	77	–	69	37
U	33	23	49	21	69	–	56
V	39	55	50	71	37	56	–

The table shows the travel time, in minutes, between seven town halls P, Q, R, S, T, U and V. Kim works at P and must visit each of the other town halls to deliver leaflets. She wishes to minimise her route.

b Find a minimum connector for the network. You must make your method clear by listing the arcs in order of selection.

c Use the minimum connector and shortcuts to find an upper bound below 220. You must list the shortcuts you use and your final route.

d Starting at P, find a nearest neighbour route and state its length.

e Find a lower bound for the length of the route by deleting P.

f Looking at your answers to **c**, **d** and **e**, use inequalities to write down the smallest interval containing the optimal solution.

5

	A	B	C	D	E	F	G
A	–	103	89	42	54	143	153
B	103	–	60	98	56	99	59
C	89	60	–	65	38	58	77
D	42	98	65	–	45	111	139
E	54	56	38	45	–	95	100
F	143	99	58	111	95	–	75
G	153	59	77	139	100	75	–

A computer supplier has outlets in seven cities A, B, C, D, E, F and G. The table shows the distances, in km, between each of these seven cities. John lives in city A and has to visit each of these cities to advise on displays. He wishes to plan a route starting and finishing at A, visiting each city and covering a minimum distance.

a Obtain a minimum spanning tree for this network explaining briefly how you applied the algorithm that you used. (Start with A and state the order in which you selected the arcs used in your tree.)

b Hence determine an initial upper bound for the length of the route travelled by John.

c Explain why the upper bound found in this way is unlikely to give the minimum route length.

d Starting from your initial upper bound and using an appropriate method, find an upper bound for the length of the route which is less than 430 km.

e By deleting city A, determine a lower bound for the length of John's route.

f Explain under what circumstances a lower bound obtained by this method might be an optimum solution.

6

	L	C	O	B	N	E
London (L)	–	80	56	120	131	200
Cambridge (C)	80	–	100	98	87	250
Oxford (O)	56	100	–	68	103	154
Birmingham (B)	120	98	68	–	54	161
Nottingham (N)	131	87	103	54	–	209
Exeter (E)	200	250	154	161	209	–

A sales representative, Sheila, has to visit clients in six cities, London, Cambridge, Oxford, Birmingham, Nottingham and Exeter. The table shows the distances, in miles, between these six cities. Sheila lives in London and plans a route starting and finishing in London. She wishes to visit each city and drive the minimum distance.

a Starting from London, use Prim's algorithm to obtain a minimum spanning tree. Show your working. State the order in which you selected the arcs and draw the tree.

b i Hence determine an initial upper bound for the length of the route planned by Sheila.

ii Starting from your initial upper bound and using shortcuts, obtain a route which is less than 660 miles.

c By deleting Exeter from the table determine a lower bound for the length of Sheila's route.

Summary of key points

1 A **walk** in a network is a finite sequence of edges such that the end vertex of one edge is the start vertex of the next.

2 A walk which visits every vertex, returning to its starting vertex, is called a **tour**.

3 The **optimal solution** is found between the **best upper bound** and **best lower bound**.

4 In the **classical travelling salesman problem** you must visit each vertex **only once**, before returning to the start.

5 In the **practical travelling salesman problem** you must visit each vertex **at least once** before returning to the start.

6 If you convert a network into a complete network of least distances, the practical and classical travelling salesman problems are the same.

7 To create a complete network of least distances, you ensure that the **triangle inequality** holds for all triangles in the network:

the longest side ⩽ the sum of the two shorter sides

8 Use the minimum spanning tree method to find an upper bound:

- find the minimum spanning tree to guarantee that each vertex is included
- double this minimum connector so that completing the cycle is guaranteed
- seek shortcuts (use some of the non-included arcs to enable you to bypass a repeat of some of the minimum spanning tree).

9 Aim to make the upper bound as low as possible to reduce the interval in which the optimal solution is contained.

10 Use the minimum spanning tree method to find a lower bound:

- remove each vertex in turn, together with its arcs
- find the residual minimum spanning tree (RMST) and its length
- add to the RMST the 'cost' of reconnecting the deleted vertex by the two shortest, distinct, arcs and note the totals
- the greater of these totals is used for the lower bound.

11 The nearest neighbour algorithm can also be used to find an upper bound:

- select each vertex in turn as a starting point
- go to the nearest vertex which has not yet been visited
- repeat until all vertices have been visited and then return to the start vertex using the shortest route
- once all the vertices have been used as the starting vertex, select the tour with the smallest length as the upper bound.

In this chapter you will:

- formulate problems as linear programming problems
- make use of slack variables
- learn how the simplex algorithm works to solve maximising linear programming problems
- solve maximising linear programming problems using simplex tableaux
- use the simplex tableau method to solve maximising linear programming problems that require integer solutions.

Further linear programming

In book D1 you studied linear programming in two variables, using a graphical method. You now move on to problems in more than two variables using an algebraic technique known as the **simplex method**.

This method is used in all types of industry and commerce from farming to oil refining, from pension fund management to production optimisation.

4.1 You are able to formulate problems as linear programming problems.

This section extends the ideas met in book D1.

Example 1

A grower specialising in cut flowers is considering cultivating four varieties of the new 'Sunlip' flowers A, B, C and D in one of his fields. He estimates the time, in hours per hectare, taken for each of four stages – sowing, thinning, picking and packing, for each variety.

For variety A sowing, thinning, picking and packing will take 3, 18, 20 and 24 hours per hectare respectively.

For variety B sowing, thinning, picking and packing will take 4, 17, 25 and 27 hours per hectare respectively.

For variety C sowing, thinning, picking and packing will take 3, 19, 26 and 28 hours per hectare respectively.

For variety D sowing, thinning, picking and packing will take 5, 16, 22 and 23 hours per hectare respectively.

For this crop he can devote up to 70 hours to sowing, 360 to thinning, 500 to picking and 550 to packing.

He estimates the total profit, in pounds per hectare as 67, 63, 71 and 75 for varieties A, B, C and D respectively.

He has up to 20 hectares to use for this crop and wishes to maximise his profit.

Formulate this as a linear programming problem. Define your variables, state your objective and write your constraints as inequalities.

It is clearer if you summarise the information in a table.

Time per hectare	A	B	C	D	Total time available
Sowing	3	4	3	5	70
Thinning	18	17	19	16	360
Picking	20	25	26	22	500
Packing	24	27	28	23	550
Profit	67	63	71	75	

FIRST define your decision variables:

Let x_A, x_B, x_C and x_D, be the number of hectares planted of variety A, B, C and D respectively.

With more than two variables we often use x_A, x_B, x_C, ... or x_1, x_2, x_3, ... instead of letters of the alphabet. x_1, x_2, x_3, ... is particularly useful when there are a large number of variables.

Next write down the objective function:

Maximise $P = 67x_A + 63x_B + 71x_C + 75x_D$

Remember that we need a word too – either maximise or minimise.

Finally write down the constraints:

Subject to:

$$3x_A + 4x_B + 3x_C + 5x_D \leqslant 70$$
$$18x_A + 17x_B + 19x_C + 16x_D \leqslant 360$$
$$20x_A + 25x_B + 26x_C + 22x_D \leqslant 500$$
$$24x_A + 27x_B + 28x_C + 23x_D \leqslant 550$$
$$x_A, x_B, x_C, x_D \geqslant 0$$

Example 2

In order to supplement his diet Andy wishes to take some and Vitatab, Weldo, Xtramin and Yestivit tablets. Amongst other ingredients, the contents of vitamins A, B, C and iron, in milligrams per tablet, are shown in the table.

	A	B	C	Iron
Vitatab	10	10	20	4
Weldo	15	20	10	5
Xtramin	25	15	15	3
Yestivit	20	15	20	2

Andy wishes to take tablets to provide him with at least 80, 30, 60 and 14 milligrams of vitamin A, B, C and iron per day.

Because of other factors Andy wants at least 25% of the tablets he takes to be Vitatab and wants to take at least twice as many Weldo as Yestivit.

The costs of the tablets are 4, 6, 12 and 7 pence per tablet. Andy wishes to minimise the cost.

Formulate this as a linear programming problem, defining your variables, stating your objective and writing your constraints as inequalities.

First define your decision variables:

Let x_1, x_2, x_3, x_4 be the number of Vitatab, Weldo, Xtramin and Yestivit tablets taken each day.

Next write down the objective function:

Minimise $C = 4x_1 + 6x_2 + 12x_3 + 7x_4$

Finally write down the constraints:

Subject to:

$$10x_1 + 15x_2 + 25x_3 + 20x_4 \geqslant 80$$
$$10x_1 + 20x_2 + 15x_3 + 15x_4 \geqslant 30$$
$$20x_1 + 10x_2 + 15x_3 + 20x_4 \geqslant 60$$
$$4x_1 + 5x_2 + 3x_3 + 2x_4 \geqslant 14$$

The amounts of each of the three vitamins and iron, give us the first four constraints.

$$x_1 \geq \frac{25}{100}(x_1 + x_2 + x_3 + x_4) \Rightarrow 3x_1 \geq x_2 + x_3 + x_4$$
$$\Rightarrow 3x_1 - x_2 - x_3 - x_4 \geq 0$$

> He wants at least 25%, $\frac{1}{4}$, of the tablets to be Vitatab.

$$x_2 \geq 2x_4 \Rightarrow x_2 - 2x_4 \geq 0$$

> He wants to take at least twice as many Weldo as Yestivit

$$x_1, x_2, x_3, x_4 \geq 0$$

> The numbers of tablets taken cannot be negative.

4.2 You are able to formulate problems as linear programming problems, making use of slack variables.

■ Inequalities can be transformed into **equations** using **slack variables** (so called because they represent the amount of slack between the total quantity and the amount being used).

Example 3

Rewrite the inequality

$$x_1 + 3x_2 + 5x_3 \leq 23,$$

as an equation, using the slack variable, r.

Introducing slack variable r, you would write

$$x_1 + 3x_2 + 5x_3 + r = 23$$

> The value of slack variable, r, tells us by how much $x_1 + 3x_2 + 5x_3$ is less than 23.

> We can formally define r as
> $$r = 23 - x_1 - 3x_2 - 5x_3$$

> A slack variable is rather like a sponge, absorbing the spare capacity so that the left hand side equals the right hand side.

Example 4

Rewrite the constraints for example 1 as equations using slack variables r, s, t and u.

$$3x_A + 4x_B + 3x_C + 5x_D \leq 70$$
$$18x_A + 17x_B + 19x_C + 16x_D \leq 360$$
$$20x_A + 25x_B + 26x_C + 22x_D \leq 500$$
$$24x_A + 27x_B + 28x_C + 23x_D \leq 550$$
$$x_A, x_B, x_C, x_D \geq 0$$

You add the four slack variables getting:

$$3x_A + 4x_B + 3x_C + 5x_D + r = 70$$
$$18x_A + 17x_B + 19x_C + 16x_D + s = 360$$
$$20x_A + 25x_B + 26x_C + 22x_D + t = 500$$
$$24x_A + 27x_B + 28x_C + 23x_D + u = 550$$
$$x_A, x_B, x_C, x_D, r, s, t, u \geqslant 0$$

Notice that you alter the non-negativity constraint too, since you now need to add r, s, t and u to the list of variables that cannot be negative. Slack variables cannot be negative.

Example 5

Rewrite the constraints for example 2 as equations using slack variables q, r, s, t, u and v.

$$10x_1 + 15x_2 + 25x_3 + 20x_4 \geqslant 80$$
$$10x_1 + 20x_2 + 15x_3 + 15x_4 \geqslant 30$$
$$20x_1 + 10x_2 + 15x_3 + 20x_4 \geqslant 60$$
$$4x_1 + 5x_2 + 3x_3 + 2x_4 \geqslant 14$$
$$3x_1 - x_2 - x_3 - x_4 \geqslant 0$$
$$x_2 - 2x_4 \geqslant 0$$
$$x_1, x_2, x_3, x_4 \geqslant 0$$

The slack variables are used to 'balance' the inequality. They should be added to the 'lighter' side, in this case the right hand side.

You should therefore write,

$$10x_1 + 15x_2 + 25x_3 + 20x_4 = 80 + q$$
$$10x_1 + 20x_2 + 15x_3 + 15x_4 = 30 + r$$

and so on, but it is more usual to collect all the variables on the same side so instead you write:

$$10x_1 + 15x_2 + 25x_3 + 20x_4 - q = 80$$
$$10x_1 + 20x_2 + 15x_3 + 15x_4 - r = 30$$
$$20x_1 + 10x_2 + 15x_3 + 20x_4 - s = 60$$
$$4x_1 + 5x_2 + 3x_3 + 2x_4 - t = 14$$
$$3x_1 - x_2 - x_3 - x_4 - u = 0$$
$$x_2 - 2x_4 - v = 0$$
$$x_1, x_2, x_3, x_4, q, r, s, t, u, v \geqslant 0$$

Exercise 4A

Formulate the following problems as linear programming problems. You must define your variables, state your objective and write your constraints as **equations**.

1. A company makes three types of metal box, round, square and rectangular. Each box has to pass through two machines to be cut and formed. The round, square and rectangular boxes need 4, 2 and 3 minutes respectively on the cutter and 2, 3 and 3 on the former. Both machines are available for 6 hours per day.

   The profit, in pence, made on each round, square and rectangular box is 12, 10 and 11 respectively. The company wishes to maximise its profit.

2. A company makes four different types of backpack, A, B, C and D. Each type A uses 2.5 units of material, needs 10 minutes of cutting time and 5 minutes of stitching time. These figures, together with those for types B, C and D are shown in the table

	A	B	C	D
Material in units	2.5	3	2	4
Cutting time in minutes	10	12	8	15
Stitching time in minutes	5	7	4	9

   There are 1400 units of material available each week, 150 hours per week available on the cutting machine and 80 hours available on the stitching machine.

   Market research says that they will sell at most 500 backpacks each week.

   The profit, in pounds, is 8, 7, 6 and 9 for types A, B, C and D respectively. The company wishes to maximise its profit.

3. The annual subscription to a bowls club is £40 for adults, £10 for children and £20 for seniors.

   The total number of members is restricted to 100.

   At most half the club must be children and at least a third must be adults.

   The club wishes to maximise its income from subscriptions.

4. Mrs Brown was rather alarmed to discover from her children at bedtime that (a week ago) they had promised she would make at least 100 small cakes for a cake sale at school the next day. Not wishing to let her children down, she puts the oven on and checks her cupboards and finds she has 3 kg of flour, 2 kg of butter and 1.5 kg of sugar, as well as other ingredients. Mrs Brown finds three cake recipes for rock cakes, fairy cakes and muffins. The recipe for rock cakes uses 220 g of flour, 100 g butter and 50 g sugar and makes 8 cakes. The recipe for fairy cakes uses 100 g each of flour, butter and sugar and makes 18 cakes. The recipe for muffins uses 250 g of flour, 50 g butter and 75 g sugar and makes 12 muffins. Each batch of rock cakes, fairy cakes and muffins take 10 minutes, 20 minutes and 15 minutes respectively to prepare.

   Mrs Brown wishes to minimise her preparation time.

5 Roma is moving house. She needs to pack all her extensive collection of china into special cardboard boxes which will be sold to her by the removal company. There are three sizes of box, small, medium and large. The small boxes have a capacity of $0.1\,\text{m}^3$ and will hold a maximum weight of $3\,\text{kg}$. The medium boxes have a capacity of $0.3\,\text{m}^3$ and will hold a maximum weight of $8\,\text{kg}$. The large boxes have a capacity of $0.7\,\text{m}^3$ and will hold a maximum weight of $18\,\text{kg}$. An expert from the removal company informs her that she should allow for at least $28\,\text{m}^3$ packing capacity and for at least $600\,\text{kg}$.

Roma decides that at least half of the boxes she uses should be small and that she should use at least twice as many medium as large.

She will be able to fill the boxes she buys and the cost of each small, medium and large box is 30p, 50p and 80p.

Roma wishes to minimise the cost of the boxes she buys.

4.3 You understand how the simplex algorithm works to solve maximising linear programming problems.

The simplex algorithm was developed by George Dantzig in 1947

Important: You will NOT be asked to use this algebraic simplex method in the examination. You will be asked to use the simplex tableau (see next section).

This section is given so that you can see the mathematical basis upon which the simplex tableau can be used.

You might be asked to relate the tableau to a diagram to demonstrate your understanding of the simplex method.

You found out in book D1 that the optimal solution to a linear programming problem lies at a vertex. You could simply find the coordinates of each vertex and test them to find the one that gives the maximum value of the objective function. This is not too difficult in a two-variable problem, but as the number of variables increases this becomes increasingly more difficult.

The simplex algorithm is more efficient. It starts at one vertex and then moves round in sequence, increasing the objective function, and just one of the variables, each time, until you reach the optimal solution.

- ■ **The simplex method allows you to:**
  - determine if a particular vertex, on the edge of the feasible region, is optimal,
  - decide which adjacent vertex you should move to in order to increase the value of the objective function.

- ■ **Slack variables are essential when using the simplex algorithm.**

You can think of the algorithm as being a set of sign posts on a treasure hunt. As you reach each vertex the algorithm tells you if you have reached the optimal solution and if not, which way you should go to get to the next sign post.

Example 6

Explain the significance of the slack variables in the graphical representation of the linear programming problem below.

Maximise $P = 3x + 2y$

Subject to:

$$5x + 7y + r = 70$$
$$10x + 3y + s = 60$$
$$x, y, r, s \geqslant 0$$

The feasible region, R is shown.

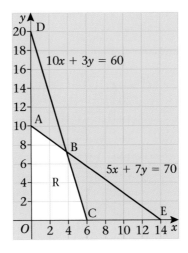

The first constraint, represented by the line ABE, where $5x + 7y = 70$, is also the line where $r = 0$.

The second constraint, represented by the line DBC, where $10x + 3y = 60$, is also the line where $s = 0$.

The four lines forming the boundaries of the feasible region can be seen as being formed by drawing the four lines $x = 0$, $y = 0$, $r = 0$ and $s = 0$. This is shown by the diagram.

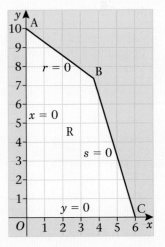

More significantly, at each vertex precisely two of our four variables and slack variables are zero.

Let's run the algebraic simplex algorithm on the two variable examples above before extending it into three dimensions.

Example **7**

This is NOT a likely examination question! You will be asked to use the simplex tableau (see next section), but not to explain how the simplex algorithm works.

Explain, in detail, how the algebraic simplex algorithm is used to solve the following problem, relating each stage to the given graph.

Maximise $P = 3x + 2y$

Subject to:

$$5x + 7y + r = 70$$
$$10x + 3y + s = 60$$
$$x, y, r, s \geqslant 0$$

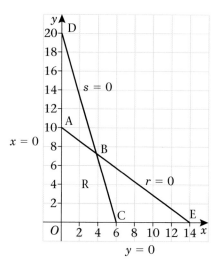

Your aim should be to **follow** this argument, you will **NOT** be asked to **produce** a similar argument in the examination.

Our inital set of equations are

$$5x + 7y + r \quad\quad = 70 \quad (1)$$
$$10x + 3y \quad\quad + s = 60 \quad (2)$$
$$P - 3x - 2y \quad\quad = 0 \quad (3)$$

Note that the objective function has been rewritten so that it is in the same 'style' as the others, in the 'usual' simultaneous equation style. The reason for this will become apparent as you move onto the simplex tableau.

The simplex method can be seen as an exercise in 'advanced' simultaneous linear equations.

Start at $O(0, 0)$. This is the vertex where $x = y = 0$

Test the objective function, $P = 3x + 2y$, at this point and get $P = 0$. This is not optimal.

Look at the objective function. You can see that if you increase x or y we will increase the objective function.

Increasing x will give a 'faster' increase – each additional x adds 3 to the value of the function.

Only increase one variable at a time so, keeping $y = 0$, increase x. (In effect this means that you move right along the x axis.)

Keep going until you hit the line $r = 0$ or the line $s = 0$. (Here you can see which one you hit first, but in more than 2 dimensions it is harder to visualise the feasible region.)

From equation (1):   If $y = 0$ and $r = 0$, then $x = 14$

From equation (2):   If $y = 0$ and $s = 0$, then $x = 6$

So you get to the vertex formed by $y = s = 0$ first, given by **equation (2)**.

Now look to see if you have reached the optimum point.

Eliminate y from the remaining equations – in this case equations (2) and (3).

Once again the reason for this will become apparent once you move into the simplex tableau.

Here are the current equations

$$5x + 7y + r \quad = 70 \quad (1)$$
$$10x + 3y \quad + s = 60 \quad (2)$$
$$P - 3x - 2y \quad = 0 \quad (3)$$

Use equation (2) because this was the equation that gave us the current vertex where $y = s = 0$

First divide **equation (2)** by 10 to make the coefficient of x, one, getting equation (5).

This gives:

$$5x + 7y + r \quad = 70 \quad (1)$$
$$x + \tfrac{3}{10}y \quad + \tfrac{1}{10}s = 6 \quad (5) \ (= (2) \div 10)$$
$$P - 3x - 2y \quad = 0 \quad (3)$$

Eliminate the x terms in equations (1) and (3). Use equation (5) to do this.

To eliminate the $+5x$ in equation (1), **subtract** 5 copies of equation (5).

To eliminate the $-3x$ in equation (3), **add** 3 copies of equation (5).

$$\tfrac{11}{2}y + r - \tfrac{1}{2}s = 40 \quad (4) \ (= (1) - 5(5))$$
$$x + \tfrac{3}{10}y \quad + \tfrac{1}{10}s = 6 \quad (5)$$
$$P \quad - \tfrac{11}{10}y \quad + \tfrac{3}{10}s = 18 \quad (6) \ (= (3) + 3(5))$$

Look at equation (6)

When $y = s = 0$, $P = 18$. So the current profit is 18.

Make P the subject of equation (6)

$$P = 18 + \tfrac{11}{10}y - \tfrac{3}{10}s$$

You can see that increasing y will increase the profit, but increasing s will decrease the profit. So you need to increase y but keep $s = 0$. This means you will travel along the line $s = 0$ in the direction that makes y increase. In this case travel along $s = 0$ to the left.

Continue along the line $s = 0$ until you hit the next vertex.

You have just left the vertex $y = s = 0$, so we will hit either $s = r = 0$ or $s = x = 0$ next.

Put these values into the current set of equations

$$\tfrac{11}{2}y + r - \tfrac{1}{2}s = 40 \quad (4)$$
$$x + \tfrac{3}{10}y \quad + \tfrac{1}{10}s = 6 \quad (5)$$
$$P \quad - \tfrac{11}{10}y \quad + \tfrac{3}{10}s = 18 \quad (6)$$

When $s = r = 0$ equation (4) gives $y = \tfrac{80}{11} = 7\tfrac{3}{11}$

When $s = x = 0$ equation (5) gives $y = 20$

You get to $y = 7\tfrac{3}{11}$ before $y = 20$, so we get to the vertex $s = r = 0$ and the value given by **equation (4)**.

Divide **equation (4)** by $\tfrac{11}{2}$ to reduce the coefficient of y to one, getting **equation (7)**

$$y + \tfrac{2}{11}r - \tfrac{1}{11}s = \tfrac{80}{11} \quad (7) = (4) \div \tfrac{11}{2}$$
$$x + \tfrac{3}{10}y \quad + \tfrac{1}{10}s = 6 \quad (5)$$
$$P \quad - \tfrac{11}{10}y \quad + \tfrac{3}{10}s = 18 \quad (6)$$

We eliminate the terms in y from the other two equations.

To eliminate $+\tfrac{3}{10}y$ from equation (5) we **subtract** $\tfrac{3}{10}$ copies of **equation (7)**.

To eliminate $-\tfrac{11}{10}y$ from equation (6) we **add** $\tfrac{11}{10}$ copies of **equation (7)**.

$$y + \tfrac{2}{11}r - \tfrac{1}{11}s = \tfrac{80}{11} \quad (7)$$
$$x \quad - \tfrac{3}{55}r + \tfrac{7}{55}s = \tfrac{42}{11} \quad (8) = (5) - \tfrac{3}{10}(7)$$
$$P \quad + \tfrac{1}{5}r + \tfrac{1}{5}s = 26 \quad (9) = (6) + \tfrac{11}{10}(7)$$

Rearranging equation (9) to make P the subject

$$P = 26 - \tfrac{1}{5}r - \tfrac{1}{5}s$$

At this vertex $r = s = 0$, so $P = 26$

You can see that if we increase r or s we will decrease the profit, so you have reached the optimal solution.

Profit = 26 when $x = \tfrac{42}{11}$, $y = \tfrac{80}{11}$, $r = 0$ and $s = 0$.

Apply this method to a problem with 3 (non-slack) variables.

Example 8

This is NOT a likely examination question! You will be asked to use the simplex tableau (see next section), but not to explain how the simplex algorithm works.

Show, in detail, the application of the algebraic simplex algorithm to the linear programming problem below.

Maximise $P = 10x + 12y + 8z$

Subject to:

$$2x + 2y \leqslant 5$$
$$5x + 3y + 4z \leqslant 15$$
$$x, y, z \geqslant 0$$

Your aim should be to **follow** this argument, you will **NOT** be asked to **produce** a similar argument in the examination.

Introducing slack variables r and s, and forming equations you get

$$2x + 2y \quad + r \quad = 5 \quad (1)$$
$$5x + 3y + 4z \quad + s = 15 \quad (2)$$
$$P - 10x - 12y - 8z \quad = 0 \quad (3)$$

Start at $(0, 0, 0)$. At this point $P = 0$, which is clearly not the maximum.

Rewrite equation (3) to make P the subject:

$$P = 10x + 12y + 8z$$

You can see that you can increase the profit by increasing x or y or z. Increase one variable at a time.

Choose to increase y, since this will give the greatest increase in the profit, and leave x and z still at zero.

As you increase y you must not leave the feasible region, however, so you must stop the first time we hit a vertex. The vertices will occur

Either when $x = z = r = 0$ or when $x = z = s = 0$

Work out the y value at each of these. The vertex with the lowest y value will be reached first.

From equation (1)   When $x = z = r = 0$, $y = \frac{5}{2}$

From equation (2)   When $x = z = s = 0$, $y = 5$

So you reach $\left(0, \frac{5}{2}, 0\right)$ first.

You now need to find out what the profit is at this point, and whether it is possible to increase it still further.

Since, at this point the values of x, z and r are all zero, and you have already increased y to its maximum, you can use equation (1) to eliminate y from both of the other equations.

Divide equation (1) by 2 to reduce the coefficient of y to one. This creates the 'pivot equation' (4)

$$x + y \qquad + \tfrac{1}{2}r \quad = \tfrac{5}{2} \qquad (4) \qquad = (1) \div 2$$

$$5x + 3y + 4z \qquad + s = 15 \qquad (2)$$

$$P - 10x - 12y - 8z \qquad = 0 \qquad (3)$$

Eliminate the y terms in the other equations, by adding or subtracting multiples of equation (4).

> To eliminate the $3y$ term from equation (2), **subtract** 3 copies of equation (4).
> To eliminate the $-12y$ term from equation (3), **add** 12 copies of equation (4).

This gives the following set of equations.

$$x + y \qquad + \tfrac{1}{2}r \quad = \tfrac{5}{2} \qquad (4)$$

$$2x \qquad + 4z - \tfrac{3}{2}r + s = \tfrac{15}{2} \qquad (5) \qquad = (2) - 3(4)$$

$$P + 2x \qquad - 8z + 6r \quad = 30 \qquad (6) \qquad = (3) + 12(4)$$

Rewriting equation (6) to make P the subject

$$P = 30 - 2x + 8z - 6r$$

> You must not increase x or r since these will decrease the profit, so we must keep x and r zero and increase z.

you see that you can still increase the profit by increasing z.

Once again, keep the other variables in that equation zero, and determine which vertex we reach first.

(In this case (equation 4) will not give a value for z – there is no z term! This speeds things up.)

From equation (5) when $x = r = s = 0$ $\qquad z = \tfrac{15}{8}$

Divide equation (5) by 4 to reduce the coefficient of z to one giving equation (7).

$$x + y \qquad + \tfrac{1}{2}r \quad = \tfrac{5}{2} \qquad (4)$$

$$\tfrac{1}{2}x \qquad + z - \tfrac{3}{8}r + \tfrac{1}{4}s = \tfrac{15}{8} \qquad (7) \qquad = (5) \div 4$$

$$P + 2x \qquad - 8z + 6r \quad = 30 \qquad (6)$$

Eliminate z from the other equations using equation (7).

There is no z term in equation (4) so this remains unchanged.

To eliminate the $-8z$ in equation (6), **add** 8 copies of equation (7).

$$x + y \quad\quad + \tfrac{1}{2}r \quad\quad = \tfrac{5}{2} \quad\quad \text{(4) (unchanged)}$$

$$\tfrac{1}{2}x \quad\quad + \quad z - \tfrac{3}{8}r + \tfrac{1}{4}s = \tfrac{15}{8} \quad\quad \text{(7)}$$

$$P + 6x \quad\quad\quad + 3r + 2s = 45 \quad\quad \text{(8)} = \text{(6)} + 8\text{(7)}$$

Make P the subject of equation (8)

$$P = 45 - 6x - 3r - 2s$$

This tells you that at the current vertex, where $x = r = s = 0$, $P = 45$

Increasing x or r or s will decrease the profit, so you cannot increase the profit further. You have found the optimal solution.

$$\text{Profit} = 45 \text{ when } x = 0, y = \tfrac{5}{2}, z = \tfrac{15}{8}, r = 0 \text{ and } s = 0.$$

Note the fact that $r = 0$ means the first constraint is at capacity, and the fact that $s = 0$ means that the second constraint is at capacity.

4.4 You can solve maximising linear programming problems using simplex tableaux.

In the examination,
- problems will be restricted to a maximum of four variables and four constraints in addition to the non-negativity constraints,
- you will only be asked to solve maximising linear problems (not minimising).

Start by solving two examples in the last section, but using simplex tableaux.

Example 9

Solve the linear programming problem in example 7 using simplex tableaux.

The word tableau is French, in French the plural of tableau is tableaux, so one tableau but three tableaux.

$$\text{Maximise } P = 3x + 2y$$

Subject to:

$$5x + 7y + r = 70$$
$$10x + 3y + s = 60$$
$$x, y, r, s \geqslant 0$$

Our initial tableau is:

Basic variable	x	y	r	s	Value
r	5	7	1	0	70
s	10	3	0	1	60
P	−3	−2	0	0	0

Compare this with our first set of equations from the algebraic solution

$$5x + 7y + r \qquad = 70 \quad (1)$$
$$10x + 3y \quad + s = 60 \quad (2)$$
$$P - 3x - 2y \qquad = 0 \quad (3)$$

If you compare these equations with the first tableau you should see that the columns marked x, y, r, s and value, just give the coefficients of those letters in the equations. You do not need to write down any of the letters at all in the tableau.

The first row shows the first constraint, the second row shows the second constraint and the final row shows the objective function.

The column marked 'basic variable' indicates the variables that are not currently at zero. Initially you start at the vertex (0, 0), so $x = y = 0$.

If $x = y = 0$ then $r = 70$ [from equation (1)] and $s = 60$ [from equation (2)]

You read across the tableau, so this initial tableau tells you that $r = 70$, $s = 60$ and $P = 0$. All other variables are zero, by definition, since they are not listed as basic variables.

Currently therefore
$$P = 0 \quad x = 0 \quad y = 0 \quad r = 70 \text{ and } s = 60$$

Now scan the objective (bottom) row of the tableau for the most negative number – in this case −3.

This gives the **pivot column** as the x column.

> In the algebraic example you found that increasing x initially was the most effective way of increasing the profit.

For each of the other rows, calculate the θ values where

$$\theta = \text{(the term in the value column)} \div \text{(the term in the pivot column)}$$

Basic variable	x	y	r	s	Value	θ values
r	5	7	1	0	70	$70 \div 5 = 14$
s	10	3	0	1	60	$60 \div 10 = 6$
P	−3	−2	0	0	0	

Select the row containing the smallest (positive) θ value. (This will become the pivot row.) In this case this is the second row.

Basic variable	x	y	r	s	Value	θ values
r	5	7	1	0	70	$70 \div 5 = 14$
s	10	3	0	1	60	$60 \div 10 = 6$
P	−3	−2	0	0	0	

We divide this row by the **pivot**, (which is the value in the pivot row and pivot column), to create the **pivot row**.

In this case we divide all the elements in row 2 by 10.

Basic variable	x	y	r	s	Value	Row operations
r	5	7	1	0	70	
x	1	$\frac{3}{10}$	0	$\frac{1}{10}$	6	R2 ÷ 10
P	−3	−2	0	0	0	

Note that the basic variable entry has also been changed. You are leaving the starting vertex where $x = y = 0$ and increasing x, so x is no longer equal to zero.

The basic variable changes to the variable in the pivot column.

We replace the s at the start of the pivot row by x, the variable at the top of the pivot column.

So there have been changes to every cell in the pivot row.

Comparing this second tableau with the second set of equations from example 7, you see that you are still matching the algebraic solution, but without having to write down all the algebra.

[The 'row operations' have been left in these equations so that you can see how they match those in the tableau.]

$$5x + 7y + r \qquad\qquad = 70 \quad (1)$$
$$x + \tfrac{3}{10}y \qquad + \tfrac{1}{10}s = 6 \quad (5) = (2) \div 10$$
$$P - 3x - 2y \qquad\qquad = 0 \quad (3)$$

Use this pivot row to eliminate x from each of the other rows.

We are aiming to get just one number 1 and make all other terms zero in the x column.

Basic variable	x	y	r	s	Value	Row operations
r	0	$\frac{11}{2}$	1	$-\frac{1}{2}$	40	R1 − 5R2
x	1	$\frac{3}{10}$	0	$\frac{1}{10}$	6	R2 ÷ 10
P	0	$-\frac{11}{10}$	0	$\frac{3}{10}$	18	R3 + 3R2

R1 − 5R2 is standard notation stating that for each entry in the row, you took the row 1 number and subtracted 5 times the row 2 entry in the same column.
So the calculations you did for row 1 were:

In column x: $5 - 5 \times 1 = 0$ \qquad In column y: $7 - 5 \times \frac{3}{10} = \frac{11}{2}$

In column r: $1 - 5 \times 0 = 1$ \qquad In column s: $0 - 5 \times \frac{1}{10} = -\frac{1}{2}$

In the value column $70 - 5 \times 6 = 40$

Similarly R3 + 3R2 states that we took the row 3 entry and added three times the corresponding row 2 entry.

So the calculations we did for row 3 were:

In column x: $-3 + 3 \times 1 = 0$ \qquad In column y: $-2 + 3 \times \frac{3}{10} = -\frac{11}{10}$

In column r: $0 + 3 \times 0 = 0$ \qquad In column s: $0 + 3 \times \frac{1}{10} = \frac{3}{10}$

In the value column $0 + 3 \times 6 = 18$

Compare this third tableau with the third set of equations

$$\frac{11}{2}y + r - \frac{1}{2}s = 40 \quad (4) \qquad = (1) - 5(5)$$
$$x + \frac{3}{10}y \qquad + \frac{1}{10}s = 6 \quad (5)$$
$$P \qquad - \frac{11}{10}y \qquad + \frac{3}{10}s = 18 \quad (6) \qquad = (3) + 3(5)$$

Now repeat the process again.

First we look for the most negative entry in the objective (bottom) row. In this case it is the $-\frac{11}{10}$ in the y column. This gives the new pivot column.

Basic variable	x	y	r	s	Value
r	0	$\frac{11}{2}$	1	$-\frac{1}{2}$	40
x	1	$\frac{3}{10}$	0	$\frac{1}{10}$	6
P	0	$-\frac{11}{10}$	0	$\frac{3}{10}$	18

Second calculate the new θ values.

Basic variable	x	y	r	s	Value	Row operations
r	0	$\frac{11}{2}$	1	$-\frac{1}{2}$	40	$40 \div \frac{11}{2} = \frac{80}{11} = 7\frac{3}{11}$
x	1	$\frac{3}{10}$	0	$\frac{1}{10}$	6	$6 \div \frac{3}{10} = 20$
P	0	$-\frac{11}{10}$	0	$\frac{3}{10}$	18	

Third we select the smallest, positive θ value. This lies in the first row, so this will become the next pivot row.

Basic variable	x	y	r	s	Value	θ values
r	0	$\frac{11}{2}$	1	$-\frac{1}{2}$	40	$40 \div \frac{11}{2} = \frac{80}{11} = 7\frac{3}{11}$
x	1	$\frac{3}{10}$	0	$\frac{1}{10}$	6	$6 \div \frac{3}{10} = 20$
P	0	$-\frac{11}{10}$	0	$\frac{3}{10}$	18	

Fourth divide the row by the pivot to create the pivot row. In this case we divide row 1 by $\frac{11}{2}$, not forgetting to change the basic variable.

Basic variable	x	y	r	s	Value	Row operations
y	0	1	$\frac{2}{11}$	$-\frac{1}{11}$	$\frac{80}{11}$	R1 $\div \frac{11}{2}$
x	1	$\frac{3}{10}$	0	$\frac{1}{10}$	6	
P	0	$-\frac{11}{10}$	0	$\frac{3}{10}$	18	

Here are the fourth set of equations determined in example 7

$$y + \frac{2}{11}r - \frac{1}{11}s = \frac{80}{11} \quad (7) \qquad = (4) \div \frac{11}{2}$$
$$x + \frac{3}{10}y \qquad + \frac{1}{10}s = 6 \quad (5)$$
$$P \qquad - \frac{11}{10}y \qquad + \frac{3}{10}s = 18 \quad (6)$$

Finally, eliminate the pivot term from the other two rows, using the pivot row to do so. In this case eliminate y from the x and P rows.

Basic variable	x	y	r	s	Value	Row operations
y	0	1	$\frac{2}{11}$	$-\frac{1}{11}$	$\frac{80}{11}$	
x	1	0	$-\frac{3}{55}$	$\frac{7}{55}$	$\frac{42}{11}$	$R2 - \frac{3}{10}R1$
P	0	0	$\frac{1}{5}$	$\frac{1}{5}$	26	$R3 + \frac{11}{10}R1$

Here are the corresponding set of equations from example 7

$$y + \tfrac{2}{11}r - \tfrac{1}{11}s = \tfrac{80}{11} \quad (7)$$
$$x \qquad -\tfrac{3}{55}r + \tfrac{7}{55}s = \tfrac{42}{11} \quad (8) \qquad = (5) - \tfrac{3}{10}(7)$$
$$P \qquad +\tfrac{1}{5}r + \tfrac{1}{5}s = 26 \quad (9) \qquad = (6) + \tfrac{11}{10}(7)$$

In the examination you might be asked to write down the equations from a given tableau.

Look along the objective row for the most negative number, but all numbers in this row are non-negative so you know that you have reached the optimal solution.

Basic variable	x	y	r	s	Value
y	0	1	$\frac{2}{11}$	$-\frac{1}{11}$	$\frac{80}{11}$
x	1	0	$-\frac{3}{55}$	$\frac{7}{55}$	$\frac{42}{11}$
P	0	0	$\frac{1}{5}$	$\frac{1}{5}$	26

Looking at the basic variable column and the value column we see that

$P = 26$, $y = \frac{80}{11}$ and $x = \frac{42}{11}$ and all other variables, and slack variables are zero.

So our full solution is

$$P = 26, x = \tfrac{42}{11}, y = \tfrac{80}{11}, r = 0, \text{ and } s = 0$$

In the examination you should state the values of P, each variable and each slack variable, as your final answer.

■ Using a simplex tableau to solve a maximising linear programming problem, where the constraints are given as equalities.

1 Draw the tableaux.
You need a basic variable column on the left, one column for each variable (including the slack variables) and a value column. You need one row for each constraint and the bottom row for the objective function.

2 Create the initial tableau.
Enter the coefficients of the variables in the appropriate column and row.

3 Look along the objective row for the most negative entry: this indicates the pivot column.

4 Calculate the θ values, for each of the constraint rows, where
θ = (the term in the value column) ÷ (the term in the pivot column)

5 Select the row with the smallest, positive θ value to become the pivot row.

6 The element in the pivot row and pivot column is the pivot.

7 Divide the row found in step 5 by the pivot, and change the basic variable at the start of the row to the variable at the top of the pivot column. This is now the pivot row.

8 Use the pivot row to eliminate the pivot's variable from the other rows.
This means that the pivot column now contains one 1 and zeros.

9 Repeat steps 3 to 8 until there are no more negative numbers in the objective row.

10 The tableau is now optimal and the non-zero values can be read off using the basic variable column and value column.

Example 10

a Use simplex tableaux to solve the linear programming problem below (from example 8).

$$\text{Maximise } P = 10x + 12y + 8z$$

Subject to:

$$2x + 2y \leqslant 5$$
$$5x + 3y + 4z \leqslant 15$$
$$x, y, z \geqslant 0$$

b Verify, using the initial equation, that your solution is feasible.

---

a Introducing slack variables r and s, and forming equations we get

$$2x + 2y \qquad + r \qquad = 5 \qquad (1)$$
$$5x + 3y + 4z \qquad + s = 15 \qquad (2)$$
$$P - 10x - 12y - 8z \qquad = 0 \qquad (3)$$

Step 2

Basic variable	x	y	z	r	s	Value
r	2	2	0	1	0	5
s	5	3	4	0	1	15
P	−10	−12	−8	0	0	0

Step 3

The most negative entry in the objective row is −12, this becomes the pivot column.

Basic variable	x	y	z	r	s	Value
r	2	2	0	1	0	5
s	5	3	4	0	1	15
P	−10	−12	−8	0	0	0

Step 4

Now calculate the θ values.

Basic variable	x	y	z	r	s	Value	θ values
r	2	2	0	1	0	5	$5 \div 2 = 2\frac{1}{2}$
s	5	3	4	0	1	15	$15 \div 3 = 5$
P	−10	−12	−8	0	0	0	

Step 5

The smallest positive θ value lies in the r row, so this will become the pivot row.

Basic variable	x	y	z	r	s	Value	θ values
r	2	2	0	1	0	5	$5 \div 2 = 2\frac{1}{2}$
s	5	3	4	0	1	15	$15 \div 3 = 5$
P	−10	−12	−8	0	0	0	

Step 6

Divide the first row by 2 and change the basic variable.

Basic variable	x	y	z	r	s	Value	Row operations
y	1	1	0	$\frac{1}{2}$	0	$2\frac{1}{2}$	R1 ÷ 2
s	5	3	4	0	1	15	R2
P	−10	−12	−8	0	0	0	R3

This is now
the pivot row.

Use this pivot row to eliminate y from the other two rows.

Basic variable	x	y	z	r	s	Value	Row operations
y	1	1	0	$\frac{1}{2}$	0	$2\frac{1}{2}$	
s	2	0	4	$-1\frac{1}{2}$	1	$7\frac{1}{2}$	R2 − 3R1
P	2	0	−8	6	0	30	R3 + 12R1

The y column now contains just one 1 (where the pivot was) and zeros.

Now look for the most negative entry in the objective row, to determine the pivot column.

Basic variable	x	y	z	r	s	Value
y	1	1	0	$\frac{1}{2}$	0	$2\frac{1}{2}$
s	2	0	4	$-1\frac{1}{2}$	1	$7\frac{1}{2}$
P	2	0	−8	6	0	30

Calculate the θ values.

Basic variable	x	y	z	r	s	Value	θ values
y	1	1	0	$\frac{1}{2}$	0	$2\frac{1}{2}$	$2\frac{1}{2} \div 0 \rightarrow \infty$
s	2	0	4	$-1\frac{1}{2}$	1	$7\frac{1}{2}$	$7\frac{1}{2} \div 4 = 1\frac{7}{8}$
P	2	0	−8	6	0	30	

This means that the next pivot row will be the second row.

Basic variable	x	y	z	r	s	Value	θ values
y	1	1	0	$\frac{1}{2}$	0	$2\frac{1}{2}$	$2\frac{1}{2} \div 0 \rightarrow \infty$
s	2	0	4	$-1\frac{1}{2}$	1	$7\frac{1}{2}$	$7\frac{1}{2} \div 4 = 1\frac{7}{8}$
P	2	0	−8	6	0	30	

Now divide all elements in the second row by 4 and change the basic variable to z.

Basic variable	x	y	z	r	s	Value	Row operations
y	1	1	0	$\frac{1}{2}$	0	$2\frac{1}{2}$	R1
z	$\frac{1}{2}$	0	1	$-\frac{3}{8}$	$\frac{1}{4}$	$1\frac{7}{8}$	R2 ÷ 4
P	2	0	−8	6	0	30	R3

Eliminate z from the other rows, using pivot row 2.

Basic variable	x	y	z	r	s	Value	Row operations
y	1	1	0	$\frac{1}{2}$	0	$2\frac{1}{2}$	R1 no change
z	$\frac{1}{2}$	0	1	$-\frac{3}{8}$	$\frac{1}{4}$	$1\frac{7}{8}$	
P	6	0	0	3	2	45	R3 + 8R2

There are no negatives in the objective (bottom) row, so our tableau is optimal. Read off the values of the basic variables y, z and P from the value column.

Basic variable	Value
y	$2\frac{1}{2}$
z	$1\frac{7}{8}$
P	45

The optimal solution is

$$P = 45, x = 0, y = 2\frac{1}{2}, z = 1\frac{7}{8}, r = 0, s = 0.$$

b $P - 10x - 12y - 8z = 0\,(1) \Rightarrow 45 - 10(0) - 12\left(2\frac{1}{2}\right) - 8\left(1\frac{7}{8}\right) = 0$

$2x + 2y + r = 5\,(2) \qquad\quad \Rightarrow 2(0) + 2\left(2\frac{1}{2}\right) + 0 = 5$

$5x + 3y + 4z + s = 15\,(3) \Rightarrow 5(0) + 3\left(2\frac{1}{2}\right) + 4\left(1\frac{7}{8}\right) + 0 = 15$

> The simplex tableau algorithm is much quicker than the algebraic simplex method. Although the algebra is there, it is hidden.

Reducing the number of tableaux

> This is optional. In the examination sufficient tableaux will be provided for the full, non-reduced solution.

In example 10 you wrote each line of each tableau twice. Once you are more comfortable with the algorithm you may be able to reduce the number of tableaux by combining all the row operations into one tableau. The important thing is to make sure that the pivot row is written down first. A complete solution of example 10 would look like this.

Example 10 solution – with reduced number of tableaux

$$2x + 2y \quad\quad + r \quad\quad = 5 \quad\quad (1)$$
$$5x + 3y + 4z \quad\quad + s = 15 \quad\quad (2)$$
$$P - 10x - 12y - 8z \quad\quad = 0 \quad\quad (3)$$

Basic variable	x	y	z	r	s	Value	θ values
r	2	2	0	1	0	5	$5 \div 2 = 2\frac{1}{2}*$
s	5	3	4	0	1	15	$15 \div 3 = 5$
P	-10	-12	-8	0	0	0	

We sometimes indicate the smallest θ value by putting *.

Basic variable	x	y	z	r	s	Value	Row operations
y	1	1	0	$\frac{1}{2}$	0	$2\frac{1}{2}$	R1 ÷ 2
s	2	0	4	$-1\frac{1}{2}$	1	$7\frac{1}{2}$	R2 − 3R1
P	2	0	-8	6	0	30	R3 + 12R1

The first row to be written in this tableau is the pivot row, row 1. The pivot row from the above tableau is then used to work out the other rows.

Basic variable	x	y	z	r	s	Value	θ values
y	1	1	0	$\frac{1}{2}$	0	$2\frac{1}{2}$	$2\frac{1}{2} \div 0 \to \infty$
s	2	0	4	$-1\frac{1}{2}$	1	$7\frac{1}{2}$	$7\frac{1}{2} \div 4 = 1\frac{7}{8}$
P	2	0	-8	6	0	30	

Basic variable	x	y	z	r	s	Value	Row operations
y	1	1	0	$\frac{1}{2}$	0	$2\frac{1}{2}$	R1 no change
z	$\frac{1}{2}$	0	1	$-\frac{3}{8}$	$\frac{1}{4}$	$1\frac{7}{8}$	R2 ÷ 4
P	6	0	0	3	2	45	R3 + 8R2

The first row to be written in this tableau is the pivot row, row 2. The pivot row from the above tableau is then used to work out the other rows.

The optimal solution is

$$P = 45, x = 0, y = 2\frac{1}{2}, z = 1\frac{7}{8}, r = 0, s = 0$$

Example 11

a Use the simplex tableau method to solve the following linear programming problem.

Maximise $P = 3x + 4y - 5z$

Subject to:

$$2x - 3y + 2z + r = 4$$
$$x + 2y + 4z + s = 8$$
$$y - z + t = 6$$
$$x, y, z, r, s, t \geqslant 0$$

b State the values of the objective function and every variable.

c Write down the equations given by your optimal tableau.

d Use the profit equation you wrote down in **c** to explain why your final tableau is optimal.

a

Basic variable	x	y	z	r	s	t	Value	θ values
r	2	−3	2	1	0	0	4	$-\frac{4}{3}$
s	1	2	4	0	1	0	8	4
t	0	1	−1	0	0	1	6	6
P	−3	−4	5	0	0	0	0	

> The most negative entry in the objective row lies in the y column, so this is the pivot column.
> Use the smallest, positive θ value, so although $-\frac{3}{4}$ is the smallest, you can not use it as a pivot.
> The smallest, positive θ value is 4, so row 2 will become the pivot row.
> The pivot is the 2 in row 2 column y, and the basic variable will change to y.

Basic variable	x	y	z	r	s	t	Value	Row operations
r	2	−3	2	1	0	0	4	R1
y	$\frac{1}{2}$	1	2	0	$\frac{1}{2}$	0	4	R2 ÷ 2
t	0	1	−1	0	0	1	6	R3
P	−3	−4	5	0	0	0	0	R4

Now divide row 2 by the pivot, 2.

Basic variable	x	y	z	r	s	t	Value	Row operations
r	$\frac{7}{2}$	0	8	1	$\frac{3}{2}$	0	16	R1 + 3R2
y	$\frac{1}{2}$	1	2	0	$\frac{1}{2}$	0	4	
t	$-\frac{1}{2}$	0	−3	0	$-\frac{1}{2}$	1	2	R3 − R2
P	−1	0	13	0	2	0	16	R4 + 4R2

Now eliminate y from the other rows. Use the pivot row, R2 in the tableau above.
To eliminate the −3 from row 1 you need to add 3 copies of row 2, so R1 + 3R2.
To eliminate the 1 from row 3 you need to subtract row 2, so R3 − R2.
To eliminate the −4 from row 4 you need to add 4 copies of row 2, so R4 + 4R2.

Basic variable	x	y	z	r	s	t	Value	θ values
r	$\frac{7}{2}$	0	8	1	$\frac{3}{2}$	0	16	$\frac{32}{7} = 4\frac{4}{7}$
y	$\frac{1}{2}$	1	2	0	$\frac{1}{2}$	0	4	8
t	$-\frac{1}{2}$	0	-3	0	$-\frac{1}{2}$	1	2	-4
P	-1	0	13	0	2	0	16	

The only negative entry in the objective row lies in the x column, so this is the pivot column.
Use the smallest, **positive** θ value, so although -4 is the smallest, you cannot use it as a pivot.
The smallest, positive θ value is $4\frac{4}{7}$, so row 1 will become the pivot row.
The pivot is the $\frac{7}{2}$ in row 1 column x.

Basic variable	x	y	z	r	s	t	Value	θ values
x	1	0	$\frac{16}{7}$	$\frac{2}{7}$	$\frac{3}{7}$	0	$\frac{32}{7}$	R1 ÷ $\frac{7}{2}$
y	$\frac{1}{2}$	1	2	0	$\frac{1}{2}$	0	4	
t	$-\frac{1}{2}$	0	-3	0	$-\frac{1}{2}$	1	2	
P	-1	0	13	0	2	0	16	

Now divide row 1 by the pivot, $\frac{7}{2}$.

Basic variable	x	y	z	r	s	t	Value	Row operations
x	1	0	$\frac{16}{7}$	$\frac{2}{7}$	$\frac{3}{7}$	0	$\frac{32}{7}$	R1 ÷ $\frac{7}{2}$
y	0	1	$\frac{6}{7}$	$-\frac{1}{7}$	$\frac{2}{7}$	0	$\frac{12}{7}$	R2 $- \frac{1}{2}$R1
t	0	0	$-\frac{13}{7}$	$\frac{1}{7}$	$-\frac{2}{7}$	1	$\frac{30}{7}$	R3 $+ \frac{1}{2}$R1
P	0	0	$\frac{107}{7}$	$\frac{2}{7}$	$\frac{17}{7}$	0	$\frac{144}{7}$	R4 $+$ R1

Eliminate x from the other rows. Use the pivot row, R1.
To eliminate the $\frac{1}{2}$ from row 2 you need to subtract $\frac{1}{2}$ of row 1, so R2 $- \frac{1}{2}$R1.
To eliminate the $-\frac{1}{2}$ from row 3 you need to add $\frac{1}{2}$ row 1, so R3 $+ \frac{1}{2}$R1.
To eliminate the -1 from row 4 you need to add row 1, so R4 $+$ R1.

There are no negatives in the objective row, so you have an optimal solution.

b $\quad P = \frac{144}{7}, x = \frac{32}{7}, y = \frac{12}{7}, z = 0, r = 0, s = 0, t = \frac{30}{7}$

Basic variable	x	y	z	r	s	t	Value	Row operations
x	1	0	$\frac{16}{7}$	$\frac{2}{7}$	$\frac{3}{7}$	0	$\frac{32}{7}$	R1 $\div \frac{7}{2}$
y	0	1	$\frac{6}{7}$	$-\frac{1}{7}$	$\frac{2}{7}$	0	$\frac{12}{7}$	R2 $- \frac{1}{2}$R1
t	0	0	$-\frac{13}{7}$	$\frac{1}{7}$	$-\frac{2}{7}$	1	$\frac{30}{7}$	R3 $+ \frac{1}{2}$R1
P	0	0	$\frac{107}{7}$	$\frac{2}{7}$	$\frac{17}{7}$	0	$\frac{144}{7}$	R4 + R1

Using the first and last columns, we read off the values of P, x, y and t. All other variables are zero.

c $\quad x + \frac{16}{7}z + \frac{2}{7}r + \frac{3}{7}s = \frac{32}{7}$

$y + \frac{6}{7}z - \frac{1}{7}r + \frac{2}{7}s = \frac{12}{7}$

$-\frac{13}{7}z + \frac{1}{7}r - \frac{2}{7}s + t = \frac{30}{7}$

$P + \frac{107}{7}z + \frac{2}{7}r + \frac{17}{7}s = \frac{144}{7}$

> The profit equation often causes difficulty during the examination. It starts '$P +$' and then the rest is given by the tableau.
> Remember: the P 'pushes in at the front' of the profit equation.

d $\quad$ If we rearrange the profit equation, making P the subject we get:

$$P = \frac{144}{7} - \frac{107}{7}z - \frac{2}{7}r - \frac{17}{7}s$$

Increasing z or r or s would decrease the profit, so the solution is optimal.

Exercise 4B

Solve the linear programming problems in questions 1 to 5 using the simplex tableau algorithm.

1 Maximise $\quad P = 5x + 6y + 4z$
Subject to
$$x + 2y + r = 6$$
$$5x + 3y + 3z + s = 24$$
$$x, y, z, r, s \geqslant 0$$

2 Maximise $\quad P = 3x + 4y + 10z$
Subject to
$$x + 2y + 2z + r = 100$$
$$x + 4z + s = 40$$
$$x, y, z, r, s \geqslant 0$$

3 Maximise $P = 3x + 5y + 2z$
Subject to

$$3x + 4y + 5z + r = 10$$
$$x + 3y + 10z + s = 5$$
$$x - 2y + t = 1$$
$$x, y, z, r, s, t \geqslant 0$$

4 Maximise $P = 3x + 6y + 32z$
Subject to

$$x + 6y + 24z + r = 672$$
$$3x + y + 24z + s = 336$$
$$x + 3y + 16z + t = 168$$
$$2x + 3y + 32z + u = 352$$
$$x, y, z, r, s, t, u \geqslant 0$$

5 Maximise $P = 4x_1 + 3x_2 + 2x_3 + 3x_4$
Subject to

$$x_1 + 4x_2 + 3x_3 + x_4 + r = 95$$
$$2x_1 + x_2 + 2x_3 + 3x_4 + s = 67$$
$$x_1 + 3x_2 + 2x_3 + 2x_4 + t = 75$$
$$3x_1 + 2x_2 + x_3 + 2x_4 + u = 72$$
$$x_1, x_2, x_3, x_4, r, s, t, u \geqslant 0$$

6 For each question 1 to 5:

a verify, using the original equations, that your solution is feasible,

b write down the final set of equations given by your optimal tableau,

c use the profit equation, written in part **b**, to explain why your solution is optimal.

4.5 You can use the simplex tableau method to solve maximising linear programming problems requiring integer solutions.

Example **12**

Solve the linear programming problem from example 9, given that we require integer solutions.

$$\text{Maximise } P = 3x + 2y$$

Subject to:

$$5x + 7y \leqslant 70$$
$$10x + 3y \leqslant 60$$
$$x, y \geqslant 0$$

In example 9 we found the following solution

$$P = 26, \qquad x = \frac{42}{11} = 3\frac{9}{11}, \qquad y = \frac{80}{11} = 7\frac{3}{11}$$

We need to test points around this optimal solution.

Point	$5x + 7y \leqslant 70$	$10x + 3y \leqslant 60$	In feasible region?	$P = 3x + 2y$
$(3, 7)$	$15 + 49 < 70$	$30 + 21 < 60$	Yes	$9 + 14 = 23$
$(3, 8)$	$15 + 56 > 70$		No	
$(4, 7)$	$20 + 49 < 70$	$40 + 21 > 60$	No	
$(4, 8)$	$20 + 56 > 70$		No	

So our best integer solution is

$$P = 23, x = 3, y = 7$$

Example 13

Solve the linear programming problem from example 10, given that we require integer solutions.

$$\text{Maximise } P = 10x + 12y + 8z$$

Subject to:

$$2x + 2y \leqslant 5$$
$$5x + 3y + 4z \leqslant 15$$
$$x, y, z \geqslant 0$$

In example 10 we found the following solution

$$P = 45, \qquad x = 0, \qquad y = 2\frac{1}{2}, \qquad z = 1\frac{7}{8}$$

We need to test points around this optimal solution.

Point	$2x + 2y \leqslant 5$	$5x + 3y + 4z \leqslant 15$	In feasible region?	$P = 10x + 12y + 8z$
$(0, 2, 1)$	$0 + 4 < 5$	$0 + 6 + 4 < 15$	Yes	$0 + 24 + 8 = 32$
$(0, 2, 2)$	$0 + 4 < 5$	$0 + 6 + 8 < 15$	Yes	$0 + 24 + 16 = 40$
$(0, 3, 1)$	$0 + 6 > 5$		No	
$(0, 3, 2)$	$0 + 6 > 5$		No	

So our best integer solution is

$$P = 40, x = 0, y = 2, z = 2$$

Mixed exercise 4C

1 In a particular factory 3 types of product, A, B and C, are made. The number of each of the products made is x, y and z respectively and P is the profit in pounds. There are two machines involved in making the products which have only a limited time available. These time limitations produce two constraints.

In the process of using the simplex algorithm the following tableau is obtained, where r and s are slack variables.

Basic variable	x	y	z	r	s	Value
z	$\frac{1}{3}$	0	1	-8	1	75
y	$\frac{2}{11}$	1	0	$\frac{17}{11}$	0	56
P	$\frac{3}{2}$	0	0	$\frac{3}{4}$	0	840

a Give one reason why this tableau can be seen to be optimal (final).

b By writing out the profit equation, or otherwise, explain why a further increase in profit is not possible under these constraints.

c From this tableau deduce
   i the maximum profit,
   ii the optimum number of type A, B and C that should be produced to maximise the profit.   **E**

2 A sweet manufacturer produces packets of orange and lemon flavoured sweets. The manufacturer can produce up to 25 000 orange sweets and up to 36 000 lemon sweets per day.

Small packets contain 5 orange and 5 lemon sweets.
Medium packets contain 8 orange and 6 lemon sweets.
Large packets contain 10 orange and 15 lemon sweets.

The manufacturer makes a profit of 14p, 20p and 30p on each of the small, medium and large packets respectively. He wishes to maximise his total daily profit.

Use x, y and z to represent the number of small, medium and large packets respectively, produced each day.

a Formulate this information as a linear programming problem, making your objective function and constraints clear. Change any inequalities to equations using r and s as slack variables.

The tableau below is obtained after one complete iteration of the simplex algorithm.

Basic variable	x	y	z	r	s	Value
r	$1\frac{2}{3}$	4	0	1	$-\frac{2}{3}$	1000
z	$\frac{1}{3}$	$\frac{2}{5}$	1	0	$\frac{1}{15}$	2400
P	-4	-8	0	0	2	72 000

b Start from this tableau and continue the simplex algorithm by increasing y, until you have either completed two complete iterations or found an optimal solution.

From your final tableau

c i write down the numbers of small, medium and large packets indicated,
ii write down the profit,
iii state whether this is an optimal solution, giving your reason.

3 Tables are to be bought for a new restaurant. The owners may buy small, medium and large tables that seat 2, 4 and 6 people respectively.

The owners require at most 20% of the total number of tables to be medium sized.
The tables cost £60, £100 and £160 respectively for small, medium and large. The owners have a budget of £2000 for buying tables.

Let the number of small, medium and large tables be x, y and z respectively.

a Write down 5 inequalities implied by the constraints. Simplify these where appropriate.

The owners wish to maximise the total seating capacity, S, of the restaurant.

b Write down the objective function for S in terms of x, y and z.

c Explain why it is not appropriate to use a graphical method to solve this problem.

It is decided to use the simplex algorithm to solve this problem.

d Show that a possible initial tableau is

Basic variable	x	y	z	r	t	Value
r	−1	4	−1	1	0	0
t	3	5	8	0	1	100
S	−2	−4	−6	0	0	0

It is decided to increase z first.

e Show that, after one complete iteration, the tableau becomes

Basic variable	x	y	z	r	t	Value
r	$-\frac{5}{8}$	$\frac{37}{8}$	0	1	$\frac{1}{8}$	$\frac{25}{2}$
z	$\frac{3}{8}$	$\frac{5}{8}$	1	0	$\frac{1}{8}$	$\frac{25}{2}$
S	$\frac{1}{4}$	$-\frac{1}{4}$	0	0	$\frac{3}{4}$	75

f Perform one further complete iteration.

g Explain how you can decide if your tableau is now final.

h Find the number of each type of table the restaurant should buy and their total cost.

4 Kuddly Pals Co. Ltd. make two types of soft toy: bears and cats. The quantity of material needed and the time taken to make each type of toy is given in the table.

Toy	Material (m²)	Time (minutes)
Bear	0.05	12
Cat	0.08	8

Each day the company can process up to $20\,m^2$ of material and there are 48 worker hours available to assemble the toys.

Let x be the number of bears made and y the number of cats made each day.

a Show that this situation can be modelled by the inequalities

$$5x + 8y \leqslant 2000,$$
$$3x + 2y \leqslant 720,$$

in addition to $x \geqslant 0, y \geqslant 0$.

The profit made on each bear is £1.50 and on each cat £1.75. Kuddly Pals Co. Ltd. wishes to maximise its daily profit.

b Set up an initial simplex tableau for this problem.

c Solve the problem using the simplex algorithm.

The diagram shows a graphical representation of the feasible region.

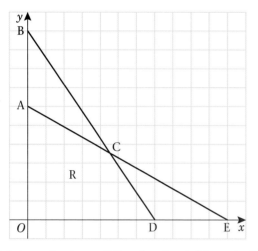

d Relate each stage of the simplex tableau to the corresponding point in the diagram.　　**E**

---

5 A clocksmith makes three types of luxury wristwatch. The mechanism for each watch is assembled by hand by a skilled watchmaker and then the complete watch is formed, weatherproofed and packaged for sale by a fitter.

The table shows the times, in minutes, for each stage of the process.

Watch type	Watchmaker	Fitter
A	54	60
B	72	36
C	36	48

The watchmaker works for a maximum of 30 hours per week and the fitter for a maximum of 25 hours per week.

Let the number of type A, B and C watches made per week be x, y and z.

a Show that the above information leads to the two inequalities

$$3x + 4y + 2z \leqslant 100,$$
$$5x + 3y + 4z \leqslant 125.$$

The profit made on type A, B and C watches is £12, £24 and £20 respectively.

b Write down an expression for the profit, P, in pounds, in terms of x, y and z.

The clocksmith wishes to maximise his weekly profit. It is decided to use the simplex algorithm to solve this problem.

c Write down the initial tableau using r and s as the slack variables.

d Increasing y first, show that after two complete iterations of the simplex algorithm the tableau becomes

Basic variable	x	y	z	r	s	Value
y	$\frac{1}{5}$	1	0	$\frac{2}{5}$	$-\frac{1}{5}$	15
z	$\frac{11}{10}$	0	1	$-\frac{3}{10}$	$\frac{2}{5}$	20
P	$\frac{74}{5}$	0	0	$\frac{18}{5}$	$\frac{16}{5}$	760

e Give a reason why this tableau is optimal (final).

f Write down the numbers of each type of watch that should be made to maximise the profit. State the maximum profit.

(E)

6 A craftworker makes three types of wooden animals for sale in wildlife parks. Each animal has to be carved and then sanded.

Each Lion takes   2 hours to carve and 25 minutes to sand.

Each Giraffe takes $2\frac{1}{2}$ hours to carve and 20 minutes to sand.

Each Elephant takes $1\frac{1}{2}$ hours to carve and 30 minutes to sand.

Each day the craftworker wishes to spend at most 8 hours carving and at most 2 hours sanding.

Let x be the number of Lions, y the number of Giraffes and z the number of Elephants he produces each day.

The craftworker makes a profit of £14 on each Lion, £12 on each Giraffe and £13 on each Elephant. He wishes to maximise his profit, P.

a Model this as a linear programming problem, simplifying your expressions so that they have integer coefficients.

It is decided to use the simplex algorithm to solve this problem.

b Explaining the purpose of r and s, show that the initial tableau can be written as:

Basic variable	x	y	z	r	s	Value
r	4	5	3	1	0	16
t	5	4	6	0	1	24
P	-14	-12	-13	0	0	0

c Choosing to increase x first, work out the next complete tableau, where the x column includes two zeros.

d Explain what this first iteration means in practical terms.

(E)

Summary of key points

1 To formulate a linear programming problem
   - define your decision variables
   - write down the objective function
   - write down the constraints.

2 Inequalities can be transformed into equations using **slack variables**.

3 The **simplex algorithm** allows you to:
   - determine if a particular vertex, on the edge of the feasible region, is optimal
   - decide which adjacent vertex you should move to in order to increase the value of the objective function.

4 To use a simplex tableau to solve a maximising linear programming problem, where the constraints are given as equalities:
   - create the initial tableau
   - look along the objective row for the most negative entry; this indicates the pivot column
   - calculate the θ values for each of the constraint rows, where
     $$\theta = \text{(the term in the value column)} \div \text{(the term in the pivot column)}$$
   - select the row with the smallest, positive, θ value to become the pivot row
   - the element in the pivot column and the pivot row is the pivot
   - divide the pivot row by the pivot, and change the basic variable at the start of the row to the variable at the top of the pivot column; this is now the pivot row
   - use the pivot row to eliminate the pivot's variable from the other rows (the pivot column now has 1 and zeros)
   - repeat until there are no more negative numbers in the objective row
   - the tableau is optimal and the non-zero values can be read off using the basic variable column and the value column.

5 When integer solutions are needed, test points around the optimal solution to find which fit the constraints and give a maximum for the objective function.

Review Exercise

Photocopy masters are available on the CD-ROM for questions marked *.

1 A theme park has four sites, A, B, C and D, on which to put kiosks. Each kiosk will sell a different type of refreshment. The income from each kiosk depends upon what it sells and where it is located. The table below shows the expected daily income, in pounds, from each kiosk at each site.

	Hot dogs and beef burgers (H)	Ice cream (I)	Popcorn, candyfloss and drinks (P)	Snacks and hot drinks (S)
Site A	267	272	276	261
Site B	264	271	278	263
Site C	267	273	275	263
Site D	261	269	274	257

Reducing rows first, use the Hungarian algorithm to determine a site for each kiosk in order to maximise the total income. State the site for each kiosk and the total expected income. You must make your method clear and show the table after each stage.

E

2 A coach company has 20 coaches. At the end of a given week, 8 coaches are at depot A, 5 coaches are at depot B and 7 coaches are at depot C. At the beginning of the next week, 4 of these coaches are required at depot D, 10 of them at depot E and 6 of them at depot F. The following table shows the distances, in miles, between the relevant depots.

	D	E	F
A	40	70	25
B	20	40	10
C	35	85	15

The company needs to move the coaches between depots at the weekend. The total mileage covered is to be a minimum.

Formulate this information as a linear programming problem.

a State clearly your decision variables.

b Write down the objective function in terms of your decision variables.

c Write down the constraints, explaining what each constraint represents. **E**

3 *

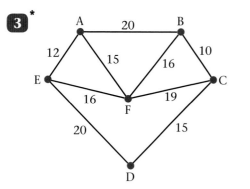

The diagram shows a network of roads connecting six villages A, B, C, D, E and F. The lengths of the roads are given in km.

a Complete the table on the worksheet, in which the entries are the shortest distances between pairs of villages. You should do this by inspection.

The table can now be taken to represent a complete network.

b Use the nearest neighbour algorithm, starting at A, on your completed table in part **a**. Obtain an upper bound to the length of a tour in this complete network, which starts and finishes at A and visits every village exactly once.

c Interpret your answer in part **b** in terms of the original network of roads connecting the six villages.

d By choosing a different vertex as your starting point, use the nearest neighbour algorithm to obtain a shorter tour than that found in part **b**. State the tour and its length. **E**

4 A manufacturing company makes 3 products X, Y and Z. The numbers of each product made are x, y and z respectively and £P is the profit. There are two machines which are available for a limited time. These time limitations produce two constraints.

In the process of using the simplex algorithm, the following tableau is obtained, where r and s are slack variables.

Basic variable	x	y	z	r	s	Value
y	0	1	$3\frac{1}{3}$	1	$-\frac{1}{3}$	1
x	1	0	-3	-1	$\frac{1}{2}$	3
P	0	0	1	1	1	33

a State how you know that this tableau is optimal (final).

b By writing out the profit equation, or otherwise, explain why a further increase in profit is not possible under these constraints.

c From this tableau, deduce
  i the maximum profit,
  ii the optimum number of X, Y and Z that should be produced to maximise the profit. **E**

5 Freezy Co. has three factories A, B and C. It supplies freezers to three shops D, E and F. The table shows the transportation cost in pounds of moving one freezer from each factory to each outlet. It also shows the number of freezers available for delivery at each factory and the number of freezers required at each shop. The total number of freezers required is equal to the total number of freezers available.

	D	E	F	Available
A	21	24	16	24
B	18	23	17	32
C	15	19	25	14
Required	20	30	20	

a Use the north-west corner rule to find an initial solution.

b Obtain improvement indices for each unused route.

c Use the stepping-stone method **once** to obtain a better solution and state its cost. **E**

6 A large room in a hotel is to be prepared for a wedding reception. The tasks that need to be carried out are:

   I   clean the room,
   II arrange the tables and chairs,
   III set the places,
   IV arrange the decorations.

The tasks need to be completed consecutively and the room must be prepared in the *least possible time*. The tasks are to be assigned to four teams of workers A, B, C and D. Each team must carry out only one task. The table below shows the times, in minutes, that each team takes to carry out each task.

	A	B	C	D
I	17	24	19	18
II	12	23	16	15
III	16	24	21	18
IV	12	24	18	14

a Use the Hungarian algorithm to determine which team should be assigned to each task. You must make your method clear and show
   i the state of the table after each stage in the algorithm,
   ii the final allocation.

b Obtain the minimum total time taken for the room to be prepared. **E**

7 A three-variable linear programming problem in x, y and z is to be solved. The objective is to maximise the profit P. The following tableau was obtained.

Basic variable	x	y	z	r	s	t	Value
s	3	0	2	0	1	$-\frac{2}{3}$	$\frac{2}{3}$
r	4	0	$\frac{7}{2}$	1	0	8	$\frac{9}{2}$
y	5	1	7	0	0	3	7
P	3	0	2	0	0	8	63

a State, giving your reason, whether this tableau represents the optimal solution.

b State the values of every variable.

c Calculate the profit made on each unit of y. **E**

8 **a** Explain the difference between the classical and practical travelling salesman problems.

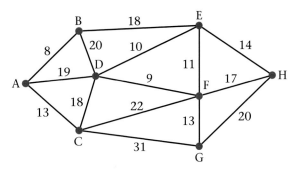

The network above shows the distances, in kilometres, between eight McBurger restaurants. An inspector from head office wishes to visit each restaurant. His route should start and finish at A, visit each restaurant at least once and cover a minimum distance.

b Obtain a minimum spanning tree for the network using Kruskal's algorithm. You should draw your tree and state the order in which the arcs were added.

c Use your answer to part **b** to determine an initial upper bound for the length of the route.

d Starting from your initial upper bound and using an appropriate method, find an upper bound which is less than 135 km. State your tour. **E**

9 In a quiz there are four individual rounds, Art, Literature, Music and Science. A team consists of four people, Donna, Hannah, Kerwin and Thomas. Each of four rounds must be answered by a different team member.

The table shows the number of points that each team member is likely to get on each individual round.

	Art	Literature	Music	Science
Donna	31	24	32	35
Kerwin	19	14	20	21
Hannah	16	10	19	22
Thomas	18	15	21	23

Use the Hungarian algorithm, reducing rows first, to obtain an allocation which maximises the total points likely to be scored in the four rounds. You must make your method clear and show the table after each stage. **(E)**

10 *

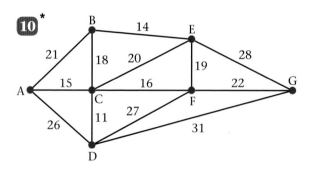

The network above shows the distances, in km, of the cables between seven electricity relay stations A, B, C, D, E, F and G. An inspector needs to visit each relay station. He wishes to travel a minimum distance, and his route must start and finish at the same station.

By deleting C, a lower bound for the length of the route is found to be 129 km.

a Find another lower bound for the length of the route by deleting F. State which is the best lower bound of the two.

b By inspection, complete the table of least distances.

The table can now be taken to represent a complete network.

c Using the nearest neighbour algorithm, starting at F, obtain an upper bound to the length of the route. State your route. **(E)**

11 Three warehouses W, X and Y supply televisions to three supermarkets J, K and L. The table gives the cost, in pounds, of transporting a television from each warehouse to each supermarket. The warehouses have stocks of 34, 57 and 25 televisions respectively, and the supermarkets require 20, 56 and 40 televisions respectively. The total cost of transporting the televisions is to be minimised.

	J	K	L
W	3	6	3
X	5	8	4
Y	2	5	7

Formulate this transportation problem as a linear programming problem. Make clear your decision variables, objective function and constraints. **(E)**

12 A manager wishes to purchase seats for a new cinema. He wishes to buy three types of seat: standard, deluxe and majestic. Let the number of standard, deluxe and majestic seats to be bought be x, y and z respectively.

He decides that the total number of deluxe and majestic seats should be at most half of the number of standard seats. The number of deluxe seats should be at least 10% and at most 20% of the total number of seats.

The number of majestic seats should be at least half of the number of deluxe seats. The total number of seats should be at least 250.

Standard, deluxe and majestic seats each cost £20, £26 and £36, respectively. The manager wishes to minimise the total cost, £C, of the seats.

Formulate this situation as a linear programming problem, simplifying your inequalities so that all coefficients are integers. **E**

13 Talkalot College holds an induction meeting for new students. The meeting consists of four talks: I (Welcome), II (Options and Facilities), III (Study Tips) and IV (Planning for Success). The four department heads, Clive, Julie, Nicky and Steve, deliver one of these talks each. The talks are delivered consecutively and there are no breaks between talks. The meeting starts at 10 a.m. and ends when all four talks have been delivered. The time, in minutes, each department head takes to deliver each talk is given in the table below.

	Talk I	Talk II	Talk III	Talk IV
Clive	12	34	28	16
Julie	13	32	36	12
Nicky	15	32	32	14
Steve	11	33	36	10

a Use the Hungarian algorithm to find the earliest time that the meeting could end. You must make your method clear and show
   i the state of the table after each stage in the algorithm,
   ii the final allocation.

b Modify the table so it could be used to find the latest time that the meeting could end. (You do not have to find this latest time.) **E**

14 The table shows the least distances, in km, between five towns, A, B, C, D and E.

	A	B	C	D	E
A	–	153	98	124	115
B	153	–	74	131	149
C	98	74	–	82	103
D	124	131	82	–	134
E	115	149	103	134	–

Nassim wishes to find an interval which contains the solution to the travelling salesman problem for this network.

a Making your method clear, find an initial upper bound starting at A and using
   i the minimum spanning tree method,
   ii the nearest neighbour algorithm.

b By deleting E, find a lower bound.

c Using your answers to parts **a** and **b**, state the smallest interval that Nassim could correctly write down. **E**

15 Three depots, F, G and H, supply petrol to three service stations, S, T and U. The table gives the cost, in pounds, of transporting 1000 litres of petrol from each depot to each service station.

	S	T	U
F	23	31	46
G	35	38	51
H	41	50	63

F, G and H have stocks of 540 000, 789 000 and 673 000 litres respectively. S, T and U require 257 000, 348 000 and 410 000 litres respectively. The total cost of transporting the petrol is to be minimised.

Formulate this problem as a linear programming problem. Make clear your decision variables, objective function and constraints. **E**

16

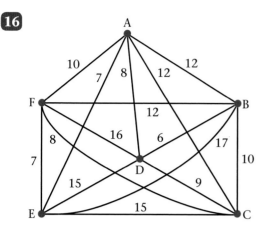

The diagram shows six towns A, B, C, D, E and F and the roads joining them. The number on each arc gives the length of that road in miles.

a By deleting vertex A, obtain a lower bound for the solution to the travelling salesman problem.

The nearest neighbour algorithm for finding a possible salesman tour is as follows:

Step 1: Let *V* be the current vertex.

Step 2: Find the nearest unvisited vertex to the current vertex, move directly to that vertex and call it the current vertex.

Step 3: Repeat step 2 until all vertices have been visited and then return directly to the start vertex.

b i Use this algorithm to find a tour starting at vertex A. State clearly the tour and give its length.

ii Starting at an appropriate vertex, use the algorithm to find a tour of shorter length. **E**

17

Warehouse / Factory	W_1	W_2	W_3	Availabilities
F_1	7	8	6	4
F_2	9	2	4	3
F_3	5	6	3	8
Requirements	2	9	4	

A manufacturer has 3 factories F_1, F_2, F_3 and 3 warehouses W_1, W_2, W_3. The table shows the cost C_{ij}, in appropriate units, of sending one unit of product from factory F_i to warehouse W_j. Also shown in the table are the number of units available at each factory F_i and the number of units required at each warehouse W_j. The total number of units available is equal to the number of units required.

a Use the north-west corner rule to obtain a possible pattern of distribution and find its cost.

b Calculate shadow costs R_i and K_j for this pattern and hence obtain improvement indices I_{ij} for each route.

c Using your answer to part **b**, explain why the pattern is optimal. **E**

18 a State the circumstances under which it is necessary to use the simplex algorithm, rather than a graphical method.

The tableau given below arose after one complete iteration of the simplex algorithm.

Basic variable	x	y	z	r	s	Value
y	$\frac{4}{5}$	1	$\frac{2}{5}$	$\frac{1}{5}$	0	$429\frac{2}{5}$
s	$2\frac{1}{5}$	0	$5\frac{3}{5}$	$-1\frac{1}{5}$	1	$1243\frac{3}{5}$
P	$-\frac{3}{5}$	0	$-\frac{4}{5}$	$1\frac{3}{5}$	0	$3435\frac{1}{5}$

b State the column that was used as the pivotal column for the first iteration.

c Perform one further complete iteration to obtain the next complete tableau.

d State the values of P, x, y and z displayed by your tableau in part **c**.

e State, giving a reason, whether your values in part **d** give the optimal solution. **E**

19 An engineering company has 4 machines available and 4 jobs to be completed. Each machine is to be assigned to one job. The time, in hours, required by each machine to complete each job is shown in the table below.

	Job 1	Job 2	Job 3	Job 4
Machine 1	14	5	8	7
Machine 2	2	12	6	5
Machine 3	7	8	3	9
Machine 4	2	4	6	10

Use the Hungarian algorithm, *reducing rows first*, to obtain the allocation of machines to jobs which minimises the total time required. State this minimum time. **E**

20 The following minimising transportation problem is to be solved.

	J	K	Supply
A	12	15	9
B	8	17	13
C	4	9	12
Demand	9	11	

a Complete the first table on the worksheet.

b Explain why an extra demand column was added to the table.

A possible north-west corner solution is:

	J	K	L
A	9	0	
B		11	2
C			12

c Explain why it was necessary to place a zero in the first row of the second column.

After three iterations of the stepping-stone method the table becomes:

	J	K	L
A		8	1
B			13
C	9	3	

d Taking the most negative improvement index as the entering square for the stepping-stone method, solve the transportation problem. You must make your shadow costs and improvement indices clear and demonstrate that your solution is optimal. **E**

21

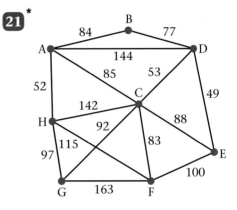

The network above shows the distances in km, along the roads between eight towns, A, B, C, D, E, F, G and H. Keith has a shop in each town and needs to visit each one. He wishes to travel a minimum distance and his route should start and finish at A.

By deleting D, a lower bound for the length of the route was found to be 586 km.
By deleting F, a lower bound for the length of the route was found to be 590 km.

a By deleting C, find another lower bound for the length of the route. State which is the best lower bound of the three, giving a reason for your answer.

b By inspection complete the table of least distances.

The table can now be taken to represent a complete network.

The nearest neighbour algorithm was used to obtain upper bounds for the length of the route:
Starting at D, an upper bound for the length of the route was found to be 838 km.
Starting at F, an upper bound for the length of the route was found to be 707 km.

c Starting at C, use the nearest neighbour algorithm to obtain another upper bound for the length of the route. State which is the best upper bound of the three, giving a reason for your answer. **E**

22 a Describe a practical problem that could be solved using the transportation algorithm.

A problem is to be solved using the transportation problem. The costs are shown in the table. The supply is from A, B and C and the demand is at d and e.

	d	e	Supply
A	5	3	45
B	4	6	35
C	2	4	40
Demand	50	60	

b Explain why it is necessary to add a third demand f.

c Use the north-west corner rule to obtain a possible pattern of distribution and find its cost.

d Calculate shadow costs and improvement indices for this pattern.

e Use the stepping-stone method once to obtain an improved solution and its cost. **E**

23 Four salespersons Ann, Brenda, Connor and Dave are to be sent to visit four companies 1, 2, 3 and 4. Each salesperson will visit exactly one company, and all companies will be visited.
Previous sales figures show that each salesperson will make sales of different values, depending on the company that they visit. These values (in £10 000s) are shown in the table below.

	1	2	3	4
Ann	26	30	30	30
Brenda	30	23	26	29
Connor	30	25	27	24
Dave	30	27	25	21

a Use the Hungarian algorithm to obtain an allocation that **maximises** the sales. You must make your method clear and show the table after each stage.

b State the value of the maximum sales.

c Show that there is a second allocation that maximises the sales. **E**

24 The manager of a car hire firm has to arrange to move cars from three garages A, B and C to three airports D, E and F so that customers can collect them. The table below shows the transportation cost of moving one car from each garage to each airport. It also shows the number of cars available in each garage and the number of cars required at each airport. The total number of cars available is equal to the total number required.

	Airport D	Airport E	Airport F	Cars available
Garage A	£20	£40	£10	6
Garage B	£20	£30	£40	5
Garage C	£10	£20	£30	8
Cars required	6	9	4	

a Use the north-west corner rule to obtain a possible pattern of distribution and find its cost.

b Calculate shadow costs for this pattern and hence obtain improvement indices for each route.

c Use the stepping-stone method to obtain an optimal solution and state its cost.

25 A chemical company makes 3 products X, Y and Z. It wishes to maximise its profit £P. The manager considers the limitations on the raw materials available and models the situation with the following linear programming problem.

Maximise $P = 3x + 6y + 4z$,
subject to $x + z \leqslant 4$,
$x + 4y + 2z \leqslant 6$,
$x + y + 2z \leqslant 12$,
$x \geqslant 0, y \geqslant 0, z \geqslant 0$,

where x, y and z are the weights, in kg, of products X, Y and Z respectively.

A possible tableau is

Basic variable	x	y	z	r	s	t	Value
r	1	0	1	1	0	0	4
s	1	4	2	0	1	0	6
t	1	1	2	0	0	1	12
P	−3	−6	−4	0	0	0	0

a Explain
  i the purpose of the variables r, s and t,
  ii the final row of the tableau.

b Solve this linear programming problem by using the simplex algorithm. Increase y for your first iteration and then increase x for your second iteration.

c Interpret your solution.

26 The table below shows the distances, in km, between six towns A, B, C, D, E and F.

	A	B	C	D	E	F
A	–	85	110	175	108	100
B	85	–	38	175	160	93
C	110	38	–	148	156	73
D	175	175	148	–	110	84
E	108	160	156	110	–	92
F	100	93	73	84	92	–

a Starting from A, use Prim's algorithm to find a minimum connector and draw the minimum spanning tree. You must make your method clear by stating the order in which the arcs are selected.

b **i** Using your answer to part **a** obtain an initial upper bound for the solution of the travelling salesman problem.
  ii Use a short cut to reduce the upper bound to a value less than 680.

c Starting by deleting F, find a lower bound for the solution of the travelling salesman problem.

27 Flatland UK Ltd makes three types of carpet, the Lincoln, the Norfolk and the Suffolk. The carpets all require units of black, green and red wool.

For each roll of carpet,
the Lincoln requires 1 unit of black, 1 of green and 3 of red,
the Norfolk requires 1 unit of black, 2 of green and 2 of red,
and the Suffolk requires 2 units of black, 1 of green and 1 of red.

There are up to 30 units of black, 40 units of green and 50 units of red available each day.
Profits of £50, £80 and £60 are made on each roll of Lincoln, Norfolk and Suffolk respectively.
Flatland UK Ltd wishes to maximise its profit.

Let the number of rolls of the Lincoln, Norfolk and Suffolk made daily be x, y and z respectively.

a Formulate the above situation as a linear programming problem, listing clearly the constraint as inequalities in their simplest form, and stating the objective function.

This problem is to be solved using the simplex algorithm. The most negative number in the profit row is taken to indicate the pivot column at each stage.

b Stating your row operations, show that after one complete iteration the tableau becomes

Basic variable	x	y	z	r	s	t	Value
r	$\frac{1}{2}$	0	$1\frac{1}{2}$	1	$-\frac{1}{2}$	0	10
y	$\frac{1}{2}$	1	$\frac{1}{2}$	0	$\frac{1}{2}$	0	20
t	2	0	0	0	-1	1	10
P	-10	0	-20	0	40	0	1600

c Explain the practical meaning of the value 10 in the top row.

d i Perform one further complete iteration of the simplex algorithm.

   ii State whether your answer to part **d i** is optimal. Give a reason for your answer.

   iii Interpret your current tableau, giving the value of each variable. **E**

28

	A	B	C	D	E	F
A	–	113	53	54	87	68
B	113	–	87	123	38	100
C	53	87	–	106	58	103
D	54	123	106	–	140	48
E	87	38	58	140	–	105
F	68	100	103	48	105	–

The table shows the distances, in km, between six towns A, B, C, D, E and F.

a Starting from A, use Prim's algorithm to find a minimum connector and draw the minimum spanning tree. You must make your method clear by stating the order in which the arcs are selected.

b i Hence form an initial upper bound for the solution to the travelling salesman problem.

   ii Use a short cut to reduce the upper bound to a value below 360.

c By deleting A, find a lower bound for the solution to the travelling salesman problem.

d Use your answers to parts **b** and **c** to make a comment on the value of the optimal solution.

e Draw a diagram to show your best route. **E**

29 Polly has a bird food stall at the local market. Each week she makes and sells three types of packs A, B and C.

Pack A contains 4 kg of bird seed, 2 suet blocks and 1 kg of peanuts.
Pack B contains 5 kg of bird seed, 1 suet block and 2 kg of peanuts.
Pack C contains 10 kg of bird seed, 4 suet blocks and 3 kg of peanuts.

Each week Polly has 140 kg of bird seed, 60 suet blocks and 60 kg of peanuts available for the packs.

The profit made on each pack of A, B and C sold is £3.50, £3.50 and £6.50 respectively. Polly sells every pack on her stall and wishes to maximise her profit, P pence.

Let x, y and z be the numbers of packs of A, B and C sold each week.

An initial simplex tableau for the above situation is

Basic variable	x	y	z	r	s	t	Value
r	4	5	10	1	0	0	140
s	2	1	4	0	1	0	60
t	1	2	3	0	0	1	60
P	-350	-350	-650	0	0	0	0

a Explain the meaning of the variables r, s and t in the context of this question.

b Perform one complete iteration of the simplex algorithm, to form a new tableau T. Take the most negative number in the profit row to indicate the pivotal column.

c State the value of every variable as given by tableau T.

d Write down the profit equation given by tableau T.

e Use your profit equation to explain why tableau T is not optimal.

Taking the most negative number in the profit row to indicate the pivotal column,

f identify clearly the location of the next pivotal element. **E**

30 A steel manufacturer has 3 factories F_1, F_2 and F_3 which can produce 35, 25 and 15 kilotonnes of steel per year, respectively. Three businesses B_1, B_2 and B_3 have annual requirements of 20, 25 and 30 kilotonnes respectively. The table below shows the cost C_{ij}, in appropriate units, of transporting one kilotonne of steel from factory F_i to business B_j.

		Business		
		B_1	B_2	B_3
Factory	F_1	10	4	11
	F_2	12	5	8
	F_3	9	6	7

The manufacturer wishes to transport the steel to the businesses at minimum total cost.

a Write down the transportation pattern obtained by using the north-west corner rule.

b Calculate all of the improvement indices I_{ij}, and hence show that this pattern is not optimal.

c Use the stepping-stone method to obtain an improved solution.

d Show that the transportation pattern obtained in part **c** is optimal and find its cost. **E**

31 A company makes three sizes of lamps, small, medium and large. The company is trying to determine how many of each size to make in a day, in order to maximise its profit. As part of the process the lamps need to be sanded, painted, dried and polished. A single machine carries out these tasks and is available 24 hours per day. A small lamp requires one hour on this machine, a medium lamp 2 hours and a large lamp 4 hours.

Let x = number of small lamps made per day,

y = number of medium lamps made per day,

z = number of large lamps made per day,

where $x \geqslant 0$, $y \geqslant 0$ and $z \geqslant 0$.

a Write the information about this machine as a constraint.

b i Re-write your constraint from part **a** using a slack variable s.

ii Explain what s means in practical terms.

Another constraint and the objective function give the following simplex tableau. The profit P is stated in euros.

Basic variable	x	y	z	r	s	Value
r	3	5	6	1	0	50
s	1	2	4	0	1	24
P	-1	-3	-4	0	0	0

c Write down the profit on each small lamp.

d Use the simplex algorithm to solve this linear programming problem.

e Explain why the solution to part **d** is not practical.

f Find a practical solution which gives a profit of 30 euros. Verify that it is feasible.　Ⓔ

32

	A	B	C	D	E	F	G	H
A	–	47	84	382	120	172	299	144
B	47	–	121	402	155	193	319	165
C	84	121	–	456	200	246	373	218
D	382	402	456	–	413	220	155	289
E	120	155	200	413	–	204	286	131
F	172	193	246	220	204	–	144	70
G	299	319	373	155	286	144	–	160
H	144	165	218	289	131	70	160	–

The table shows the distances, in miles, between some cities. A politician has to visit each city, starting and finishing at A. She wishes to minimise her total travelling distance.

a Find a minimum spanning tree for this network.

b Hence find an upper bound for this problem.

c Reduce this upper bound to a value below 1400 by using 'short cuts'.

d By deleting D find a lower bound for the distance to be travelled.

e Explain why the method used in part **d** will always give a lower bound for the distance to be travelled in any such network.　Ⓔ

33 A carpenter makes small, medium and large chests of drawers. The small size requires $2\frac{1}{2}$ m of board, the medium size 10 m of board and the large size 15 m of board. The times required to produce a small chest, a medium chest and a large chest are 10 hours, 20 hours and 50 hours respectively. In a given year there are 300 m of board available and 1000 production hours available.

Let the number of small, medium and large chests made in the year be x, y and z respectively.

a Show that the above information leads to the inequalities

$$x + 4y + 6z \leqslant 120,$$
$$x + 2y + 5z \leqslant 100.$$

The profits made on small, medium and large chests are £10, £20 and £28 respectively.

b Write down an expression for the profit £P in terms of x, y and z.

The carpenter wishes to maximise his profit. The simplex algorithm is to be used to solve this problem.

c Write down the initial tableau using r and s as slack variables.

d Use two iterations of the simplex algorithm to obtain the following tableau. In the first iteration you should increase y.

Basic variable	x	y	z	r	s	Value
y	0	1	$\frac{1}{2}$	$\frac{1}{2}$	$-\frac{1}{2}$	10
x	1	0	4	-1	2	80
P	0	0	22	0	10	1000

e Give a reason why this tableau is optimal.

f Write down the number of each type of chest that should be made to maximise the profit. State the maximum profit.　Ⓔ

34

	A	B	C	D	E	F	G
A	–	55	125	160	135	65	95
B	55	–	82	135	140	100	83
C	125	82	–	85	120	140	76
D	160	135	85	–	65	132	63
E	135	140	120	65	–	90	55
F	65	100	140	132	90	–	75
G	95	83	76	63	55	75	–

A retailer has shops in seven cities A, B, C, D, E, F and G. The table above shows the distances, in km, between each of these seven cities. Susie lives in city A and has to visit each of the shops. She wishes to plan a route starting and finishing at A and covering a minimum distance.

a Starting at A, use an algorithm to find a minimum spanning tree for this network. State the order in which you added vertices to the tree and draw your final tree. Explain briefly how you applied the algorithm.

b Hence determine an initial upper bound for the length of Susie's route.

c Starting from your initial upper bound, obtain an upper bound for the route which is less than 635 km. State the route which has a length equal to your new upper bound and cities which are visited more than once.

d Obtain the minimum spanning tree for the reduced graph produced by deleting the vertex G and all edges joined to it. Draw the tree.

e Hence obtain a lower bound for the length of Susie's route.

f Using your solution to part **d**, obtain a route of length less than 500 km which visits each vertex exactly once. *E*

35 T42 Co. Ltd produces three different blends of tea, Morning, Afternoon and Evening. The teas must be processed, blended and then packed for distribution.

The table below shows the time taken, in hours, for each stage of the production of a tonne of tea. It also shows the profit, in hundreds of pounds, on each tonne.

	Processing	Blending	Packing	Profit (£100)
Morning blend	3	1	2	4
Afternoon blend	2	3	4	5
Evening blend	4	2	3	3

The total times available each week for processing, blending and packing are 35, 20 and 24 hours respectively. T42 Co. Ltd wishes to maximise the weekly profit.

Let x, y and z be the number of tonnes of Morning, Afternoon and Evening blend produced each week.

a Formulate the above situation as a linear programming problem, listing clearly the objective function, and the constraints as inequalities.

An initial simplex tableau for the above situation is

Basic variable	x	y	z	r	s	t	Value
r	3	2	4	1	0	0	35
s	1	3	2	0	1	0	20
t	2	4	3	0	0	1	24
P	−4	−5	−3	0	0	0	0

b Solve this linear programming problem using the simplex algorithm. Take the most negative number in the profit row to indicate the pivot column at each stage.

T42 Co. Ltd wishes to increase its profit further and is prepared to increase the time available for processing or blending or packing or any two of these three.

c Use your answer to part **b** to advise the company as to which stage(s) should be allocated increased time. *E*

5

In this chapter you will:

- learn about two-person games and pay-off matrices
- learn what is meant by play safe strategies and determine the play safe strategy for each player
- learn what is meant by a zero-sum game
- be able to reduce a pay-off matrix using dominance arguments
- be able to determine the optimal mixed strategy for a game with no stable solution, for the player with two choices in a 2 × 3 or 3 × 2 game, and for the player with three choices in a 2 × 3 or 3 × 3 game
- be able to convert 2 × 3, 3 × 2 and 3 × 3 games into linear programming problems.

Game theory

Film makers compete with each other for customers.

If one maker runs a substantial advertising campaign, special offer, or competition, this is likely to have consequences, not only for their own revenue, but will also have an impact on the revenue of their competitors.

Game theory enables you to analyse decision making between two or more competitors.

5.1 You know about two-person games and pay-off matrices.

The best known classic example of game theory is the Prisoner's Dilemma which, in one version, goes as follows.

Two men are caught trying to spend a large number of forged £50 notes and are arrested by the police. The inspector in charge of the investigation is convinced that these two men are not only guilty of trying to spend forged notes but are also the counterfeiters. He has no evidence that will stand up in court at present, so he puts the men in different rooms and makes the same proposition **separately** to each of the arrested men.

'If **neither** of you confesses to being a counterfeiter, then we will charge both of you with attempting to pass forged notes – I expect you will get about one year for that crime.

Should **both** of you confess to being forgers, than we will do our best to get a lenient sentence – I would expect about four years.

However, if **only** you confesses to forgery, then we will get you a free pardon, but we will throw the book at your fellow prisoner – and I expect he will get about 10 years'.

This is an example of a two-person game.

■ A **two person game** is one in which only two parties can play.

If he *does not* confess and his fellow prisoner *does not* confess either, he might get one year in prison. However, if his fellow prisoner *does* confess, he might get 10 years.

So if he *does not* confess the best that can happen is prison for one year, the worst is prison for 10 years.

If he confesses he might get a free pardon if his fellow prisoner does not confess, but he might get four years if his fellow prisoner also confesses.

So if he confesses the best that can happen is a free pardon and the worst four years.

Summarise this in the table.

	B confesses	B does not confess	Worst outcome for A
A confesses	$(-4, -4)$	$(0, -10)$	-4 (four years in prison)
A does not confess	$(-10, 0)$	$(-1, -1)$	-10 (10 years in prison)

■ The table gives the outcomes as ordered pairs (A, B). It is called a **pay-off matrix**.

The worst outcome for each of A's choices has been added to the pay-off matrix.

If you look at the option where the worst outcome is minimised, you can see that A should choose to confess.

Of course, B has exactly the same choices and so he should also choose to confess.

This means that both prisoners confess, they get four years each and, the story goes on, the inspector gets a promotion.

5.2 You should understand what is meant by play safe strategies.

■ When **playing safe** each player looks for the worst that could happen if he makes each choice in turn. He then picks the choice that results in the least worst option.

The minimum value in each row (for player A) and each column (for player B) is listed at the end of the row/column. You then select the maximum of these minimums to find out which option the player should choose.

Example 1

In this pay-off matrix player A has a choice of four options, player B three. The outcomes are given as ordered pairs, (A's winnings, B's winnings). Determine the play safe strategy for each player.

	B plays 1	B plays 2	B plays 3
A plays 1	(8, 2)	(0, 9)	(7, 3)
A plays 2	(3, 6)	(9, 0)	(2, 7)
A plays 3	(1, 7)	(6, 4)	(8, 1)
A plays 4	(4, 2)	(4, 6)	(5, 1)

If A chooses option 1 we say that **A plays 1**.

Record the worst outcomes for each decision for each player.

	B plays 1	B plays 2	B plays 3	Worst outcome for A
A plays 1	(8, 2)	(0, 9)	(7, 3)	0
A plays 2	(3, 6)	(9, 0)	(2, 7)	2
A plays 3	(1, 7)	(6, 4)	(8, 1)	1
A plays 4	(4, 2)	(4, 6)	(5, 1)	4
Worst outcome for B	2	0	1	

For player A
If he chooses option 1 his worst outcome is a win of 0.
If he chooses option 2 his worst outcome is a win of 2.
If he chooses option 3 his worst outcome is a win of 1.
If he chooses option 4 his worst outcome is a win of 4.

For player B.
If he chooses option 1 his worst outcome is a win of 2.
If he chooses option 2 his worst outcome is a win of 0.
If he chooses option 3 his worst outcome is a win of 1.

Look at A's worst outcomes and choose the option that gives him the best result.
In this case **A should play 4**.
Look at B's worst outcomes. **B should play 1**.
By playing safe the outcome is (4, 2) — A wins 4 and B wins 2.

5.3 You should understand what is meant by a zero-sum game.

In example 1 it would pay the players to **collaborate**. For example, if A plays 3 and B plays 2 the result is (6, 4) and both players increase their winnings.

Also, if the game were played many times, it would pay each player to vary their strategy. For example if A knew that B would play 1, he could play 1 himself on occasion, leading to a win of 8. Similarly, if B were sure that A would play 4, he could play 2 on occasion, leading to a win of 6.

> Who pays out all the winnings?

It is much more likely that there is no external banker and that A's gain is covered by B's loss and vice versa. So a more plausible game could be

	B plays 1	B plays 2	B plays 3
A plays 1	(3, −3)	(−4, 4)	(2, −2)
A plays 2	(−1, 1)	(4, −4)	(−2, 2)
A plays 3	(−3, 3)	(1, −1)	(4, −4)
A plays 4	(1, −1)	(−1, 1)	(1, −1)

■ The two entries in each cell add up to zero, and so games like this are referred to as **zero-sum games**.

Since the two entries are always the negative of each other you only need to write A's winnings, since this also represents the negative of B's winnings.

Rewrite the above pay-off matrix as

$$\begin{bmatrix} 3 & -4 & 2 \\ -1 & 4 & -2 \\ -3 & 1 & 4 \\ 1 & -1 & 1 \end{bmatrix}$$

> Remember that these numbers represent A's winnings and that B's winnings are the negative of each number in the table.

■ A pay-off matrix is always written from the row player's (A's) point of view unless you are told otherwise.

In zero-sum games collaboration is not beneficial, but a varying strategy may still be useful if the game is to be played several times.

5.4 You should be able to determine the play safe strategy for each player.

■ The play-safe strategies are:
> For player A (rows) the row maximin
> For player B (columns) the column minimax.

> The maximin means find the maximum of the minimum values.
> The minimax means find the minimum of the maximum values.

Example 2

Determine the play safe strategy for both players, A and B, for the game below.

$$\begin{vmatrix} 3 & -4 & 2 \\ -1 & 4 & -2 \\ -3 & 1 & 4 \\ 1 & -1 & 1 \end{vmatrix}$$

To determine player A's strategy.

Record the minimum number in each row and choose the row giving the maximum of these minimum.

			Row minimum
3	-4	2	-4
-1	4	-2	-2
-3	1	4	-3
1	-1	1	-1

The maximum of these minimum (**maximin**) is given by row 4. **So A plays 4.**

To determine B's play-safe strategy.

Record the maximum number in each column and choose the column giving the minimum of these maximum.

	3	-4	2
	-1	4	-2
	-3	1	4
	1	-1	1
Column maximum	3	4	4

> Remember that the pay-off matrix shows the game from A's point of view.
> So a 3 means −3 from B's point of view. B's worst outcome in each column will be given by the biggest number, since this is A's biggest winning number and therefore B's biggest loss, and you want the smallest of these.

The minimum of these maximum (**minimax**) is given by column 1. **So B plays 1.**

5.5 You should understand what is meant by a stable solution (saddle point).

You have already seen that collaboration is not advantageous in a zero-sum game, but you need to consider playing a **mixed strategy**.

Example 3

Consider the following pay-off matrix.

$$\begin{vmatrix} 4 & -1 & 2 & 3 \\ 4 & 6 & 3 & 7 \\ 1 & 2 & -2 & 4 \end{vmatrix}$$

Demonstrate that there is a stable solution to this game, explaining your reasoning carefully. State the best strategy for each player and the saddle point.

Look for the play safe strategies.

					Row minimum
4	−1	2	3		−1
4	6	3	7		3
1	2	−2	4		−2
Column maximum	4	6	3	7	

To remember the headings think of a column of roman soldiers: 'Ro**min** colu**max**'.

The row maximin is 3, so A should play 2.
The column minimax is 3, so B should play 3.
This results in a win for A of 3.

You now need to decide if it is worth A or B changing their strategy.

Go through A's reasoning.
A assumes that B will play column 3.
If A looks at her options given that B plays column 3, we can see that if she plays row 1 she only wins 2 and if she plays row 3 she will lose 2.
So it would not be sensible for A to change her strategy.

Go through B's reasoning.
B assumes that A will play row 2.
If B looks at his options given this, we can see that if he plays column 1 he loses 4, column 2 he loses 6 and column 4 he loses 7. All of these are worse than the current situation where B is losing 3.
So it would not be sensible for B to change his strategy.

Thus, even if the game were to be played many times, neither player has any incentive to change the play safe strategy.

This game is said to have a **stable solution** (or is in **equilibrium**).
The **value of the game** to player A is 3,
The **value of the game** to player B is −3.
The **saddle point** is A2, B3.

Example 4

Determine whether the pay-off matrix in example 2 has a stable solution. Explain your reasoning carefully.

$$\begin{bmatrix} 3 & -4 & 2 \\ -1 & 4 & -2 \\ -3 & 1 & 4 \\ 1 & -1 & 1 \end{bmatrix}$$

In example 2, you found that, if A and B both play safe, A should play row 4, and B should play column 1.

Go through A's reasoning.

3	−4	2
−1	4	−2
−3	1	4
1	−1	1

A has solved the problem, and assumes that B will play safe and play column 1.

A would then win 1.

However, if A plays row 1 she will win 3, if she plays row 2 she loses 1, if she plays row 3 she loses 3.

Thus, if the game were to be played many times, it would pay A to play a mixed strategy and play row 1 some of the time.

> Of course, if A does this every time, B will also change his strategy, so A must develop a mixed strategy (see Section 5.7).

Go through B's reasoning.

3	−4	2
−1	4	−2
−3	1	4
1	−1	1

B assumes that A will play safe and play row 4.

If B plays column 1 he loses 1. If B plays column 2 he wins 1. If he plays column 3 he still loses 1.

Thus, if the game were to be played many times, it would pay B to play a mixed strategy and play column 2 some of the time.

Hence the game has no stable solution.

■ **In a zero-sum game there will be a stable solution if and only if**

the row maximin = the column minimax

This helpful theorem saves us doing those reasoning arguments.

Proof of the stable solution theorem.

This is not needed for the examination.

If you define

V(A) = row maximin element (i.e. the element that gives A's play safe winnings)

V(B) = − column minimax element (i.e. the element that gives B's play safe winnings)

you can prove two theorems.

Theorem 1: For any zero-sum game V(A) + V(B) ≤ 0

If both players play safe A wins at least V(A) and B wins at least V(B) so the total winnings are at least V(A) + V(B). However, in a zero-sum game the total winnings are always zero. Hence V(A) + V(B) cannot exceed zero.

Theorem 2: A zero-sum game has a stable solution if and only if V(A) + V(B) = 0

The 'if' bit (If V(A) + V(B) = 0 then the game is stable)

Since V(A) is the row maximin, there is a row, say the Rth, in which the smallest element is V(A). (A will therefore be playing safe by playing row R.)

Similarly since V(B) is the column minimax, there is a column, say the Sth, in which the largest entry is $-$V(B). (B will therefore be playing safe by playing column S.)

The value of the game will be given by the (r, s)th element.

Now you know that $(r, s) \geq$ V(A) since V(A) is the smallest number in row R.

Also you know that $(r, s) \leq -$V(B) since $-$V(B) is the biggest number in column S.

Thus

$$V(A) \leq (r, s) \leq -V(B)$$

but you are told that V(A) + V(B) = 0, hence V(A) = $-$V(B), hence

$$V(A) \leq (r, s) \leq V(A)$$

Hence $(r, s) =$ V(A) $= -$V(B)

This means that (r, s) is the element which is the smallest in row R and biggest in column S.

A assumes that B will play column S. There is no profit in A changing rows since (r, s) gives the biggest element in column S. So A will always play row R.

B assumes that A will play row R. There is no profit in B changing columns since (r, s) gives the smallest element in row R. So B will always play column S.

Thus the game is stable and (r, s) is the saddle point.

The 'only if' bit (If there is a stable solution then V(A) + V(B) = 0)

If the game is stable there is a saddle point (or points) which gives that game's value. Since you are looking at a zero-sum game you know that this must be zero.

Example 5

Consider the following pay-off matrix

	B plays 1	B plays 2
A plays 1	4	-2
A plays 2	-5	3

a Determine the play safe strategy for each player.
b Verify that there is no stable solution for this game.
c State the value of the game for A if both players play safe.
d State the value of the game for B if both players play safe.
e Determine the pay-off matrix for B.

a

	B plays 1	B plays 2	Row minimum
A plays 1	4	−2	−2
A plays 2	−5	3	−5
Column maximum	4	3	

We first record the row minimums and column maximums. The row maximin is −2 and the column minimax is 3,

The play safe solution is (1, 2) i.e. A plays row 1 and B plays column 2.

b Since −2 ≠ 3 the row maximin ≠ column minimax. There is not a stable solution.

> So both A and B could profit by not playing safe.

c If A and B play safe, A plays 1 and B plays 2.

	B plays 1	B plays 2	Row minimum
A plays 1	4	−2	−2
A plays 2	−5	3	−5
Column maximum	4	3	

The value of the game for A is −2

d The value of the game for B is 2

> If A 'wins' − 2 each time they both play safe, then B will win 2 each time.

e Make B the 'row' player and change all the signs.

	A plays 1	A plays 2
B plays 1	−4	5
B plays 2	2	−3

> For those who have studied matrices, you obtain the pay-off matrix for the other player by transposing the matrix and then multiplying each term by −1.

Exercise 5A

1 A two person zero-sum game is represented by the following pay-off matrix for player A.

	B plays 1	B plays 2	B plays 3
A plays 1	3	2	3
A plays 2	−2	1	3
A plays 3	4	2	1

a Determine the play safe strategy for each player.

b Verify that there is a stable solution for this game and determine the saddle point.

2 Robert and Steve play a zero-sum game. This game is represented by the following pay-off matrix for Robert.

	Steve plays 1	Steve plays 2	Steve plays 3	Steve plays 4
Robert plays 1	−2	−1	−3	1
Robert plays 2	2	3	1	−2
Robert plays 3	1	1	−1	3

   a Determine the play safe strategy for each player.
   b Verify that there is no stable solution for this game.

3 A two person zero-sum game is represented by the following pay-off matrix for player A.

	B plays 1	B plays 2	B plays 3
A plays 1	−3	−2	2
A plays 2	−1	−1	3
A plays 3	4	−3	1
A plays 4	3	−1	−1

   a Determine the play safe strategy for each player.
   b Verify that there is a stable solution for this game and determine the saddle points.
   c State the value of the game to player A.

4 Claire and David play a two person zero-sum game, which is represented by the following pay-off matrix for Claire.

	D plays 1	D plays 2	D plays 3	D plays 4
C plays 1	7	2	−3	5
C plays 2	4	−1	1	3
C plays 3	−2	5	2	−1
C plays 4	3	−3	−4	2

   a Determine the play safe strategy for each player.
   b Verify that there is no stable solution for this game.
   c State the value of the game for Claire if both players play safe.
   d State the value of the game for David if both players play safe.
   e Determine the pay-off matrix for David.

5 Hilary and Denis play a two person zero-sum game, which is represented by the following pay-off matrix for Hilary.

	D plays 1	D plays 2	D plays 3	D plays 4	D plays 5
H plays 1	2	1	0	0	2
H plays 2	4	0	0	0	2
H plays 3	1	4	−1	−1	3
H plays 4	1	1	−1	−2	0
H plays 5	0	−2	−3	−3	−1

a Determine the play safe strategy for each player.

b Verify that there is a stable solution for this game and state the saddle points.

c State the value of the game for Hilary if both players play safe.

d State the value of the game for Denis if both players play safe.

e Determine the pay-off matrix for Denis.

5.6 You should be able to reduce a pay-off matrix using dominance arguments.

■ If one 'named' row (or column) is *always* a better option than another 'named' row (or column) then the better option **dominates**.

■ The 'named' row (or column) that is dominated can be deleted to reduce the pay-off matrix.

Example 6

A and B play a zero-sum game, represented by the pay-off matrix below.

	B plays 1	B plays 2
A plays 1	7	1
A plays 2	-3	6
A plays 3	4	0

This is a 3 × 2 game since player A has a choice of three rows, and B a choice of two columns.

Explain why the 3 × 2 game can be reduced to a 2 × 2 game.

	B plays 1	B plays 2
A plays 1	7	1
A plays 2	-3	6
A plays 3	4	0

If you look carefully at A's row 1 and row 3 choices you can see that A would never choose to play row 3, no matter which column B is playing row 1 is a better option.

Row 1 **dominates** row 3, since $7 \geqslant 4$ and $1 \geqslant 0$.

This means that, for player A, no matter what player B does, row 1 is **always** a better option than row 3.

	B plays 1	B plays 2
A plays 1	7	1
A plays 2	-3	6
A plays 3	4	0

You can only use domination arguments to delete a 'named' row (or column) if **one** other 'named' row (column) is **always** better.

Thus you can delete row 3 and reduce the game to a 2 × 2, game.

	B plays 1	B plays 2
A plays 1	7	1
A plays 2	-3	6

Example **7**

A and B play a zero-sum game, represented by the pay-off matrix below.

	B plays 1	B plays 2	B plays 3
A plays 1	7	1	−2
A plays 2	−3	6	1

Explain why the 2 × 3 game can be reduced to a 2 × 2 game.

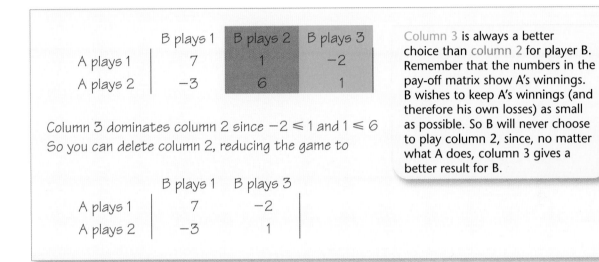

	B plays 1	B plays 2	B plays 3
A plays 1	7	1	−2
A plays 2	−3	6	1

Column 3 dominates column 2 since −2 ⩽ 1 and 1 ⩽ 6
So you can delete column 2, reducing the game to

	B plays 1	B plays 3
A plays 1	7	−2
A plays 2	−3	1

Column 3 is always a better choice than column 2 for player B. Remember that the numbers in the pay-off matrix show A's winnings. B wishes to keep A's winnings (and therefore his own losses) as small as possible. So B will never choose to play column 2, since, no matter what A does, column 3 gives a better result for B.

5.7 **You should be able to determine the optimal mixed strategy for a game with no stable solution.**

■ You should **always** check for a stable solution first. You should only consider a mixed strategy if there is no stable solution.

Imagine that A and B are playing the game many times. As you have already seen (see page 140), it would sometimes pay A and B to vary their choice. Once one player plays a mixed strategy it will always pay the other to play a mixed strategy too.

(In fact, it is possible to play the game just once and still use a mixed strategy – all will be revealed later!)

■ You work out the expected winnings for each player, using probabilities, and hence advise them on their best strategy.

■ To determine the optimal strategy, form probability equations and graph them.

■ The intersection point gives you the information you need to work out the probabilities and the value of the game.

Example 8

A and B play a zero-sum game which is represented by the following pay-off matrix for A.

	B plays 1	B plays 2
A plays 1	4	−2
A plays 2	−5	3

a Verify that A and B should play mixed strategies.

b Determine the mixed strategy each player should adopt, and the value of the game to each player.

a

	B plays 1	B plays 2	Row minimum
A plays 1	4	−2	−2
A plays 2	−5	3	−5
Column maximum	4	3	

The play safe solution is (1, 2) i.e. A plays row 1 and B plays column 2.
The row maximin is −2 and the column minimax is 3, so checking for a stable solution

$$-2 \neq 3$$

hence there is not a stable solution.
Thus both A and B could profit by not playing safe.

> A loses 2 with his play safe choice. He assumes B will play safe and play column 2 so it would pay A to sometimes move to row 2 to win 3.
>
> B wins 2 with his play safe choice. If he assumes A will play row 1 it would not pay him to move. But if A is going to play a mixed strategy to increase his winnings, any increase of A's winnings is a decrease in B's winnings (it is a zero-sum game).
>
> Also if A is going to sometimes play row 2, if B could play column 1 at the same time he would win 5 instead of 2.
>
> Thus it would pay B also to play a mixed strategy.

b **For player A**

Let A play: row 1 with a probability p,
and therefore row 2 with
a probability of $1 - p$.

> In game theory, not playing is not an option! So A **must** play either row 1 or row 2, hence the probabilities **must** add up to 1. If A plays row 1 with probability p, then the probability A does not play row 1 (and so plays row 2) is $1 - p$.

A knows that B has two options:

If B plays column 1   A's expected winnings are $V(A) = 4p - 5(1 - p) = 9p - 5$

If B plays column 2   A's expected winnings are $V(A) = -2p + 3(1 - p) = 3 - 5p$

We can illustrate these options by a graph.

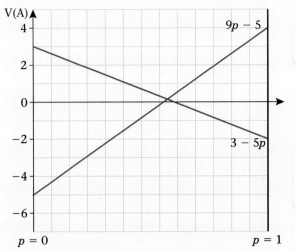

Notice that the p-axis only exists between 0 and 1. This is because it is a probability.

We usually draw axes in the form of a 'rugby post', with two vertical lines. This gives us a focus for each end of the line, when $p = 0$ and when $p = 1$.

A will maximise his winnings if he chooses p such that his minimum winnings are as high as possible. In this case it is easy to see that A should choose p such that

$$9p - 5 = 3 - 5p$$
$$14p = 8$$
$$p = \tfrac{4}{7}$$

A should play row 1 with a probability of $\tfrac{4}{7}$, and row 2 with a probability of $\tfrac{3}{7}$.

The value of the game for A, V(A), is $\tfrac{1}{7}$.

A must, of course, play randomly.

We can calculate the value of the game by substituting the value of p into either of the two equations we used to find its value.

For player B

Similarly let B play column 1 with a probability of q and therefore column 2 with a probability of $1 - q$.

B knows that A has two options:

If A plays row 1, B's expected winnings are
$$V(B) = -[4q - 2(1 - q)] = 2 - 6q$$

If A plays row 2, B's expected winnings are
$$V(B) = -[-5q + 3(1 - q)] = 8q - 3$$

Notice the minus signs outside the square brackets. The game is shown from player A's point of view, so to find B's winnings you need to change the signs.

Again we can use a graph to illustrate this information.

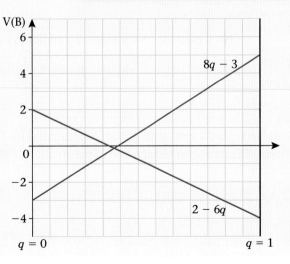

Finding the intersection gives $2 - 6q = 8q - 3$ and hence $q = \frac{5}{14}$.

Thus B should play column 1 with a probability of $\frac{5}{14}$ and column 2 with a probability of $\frac{9}{14}$.

We substitute q into either of the two equations we used to calculate the intersection point to get the value of V(B).

The value of the game for B, V(B), is $-\frac{1}{7}$.

B must play randomly.

The value of the game to A was $\frac{1}{7}$ and the value of the game to B is $-\frac{1}{7}$. This is to be expected since it is a zero-sum game. The sum of the values of the game to each player must be zero. A's winnings and B's winnings must sum to zero.

Exercise 5B

1

	Freya plays 1	Freya plays 2
Ellie plays 1	1	−5
Ellie plays 2	−1	6
Ellie plays 3	3	−3

Ellie and Freya play a zero-sum game, represented by the pay-off matrix for Ellie shown above. Use dominance to reduce the game to a 2 × 2 game. You must make your reasoning clear.

2

	Harry plays 1	Harry plays 2	Harry plays 3
Doug plays 1	−5	2	−1
Doug plays 2	2	−3	−6

Doug and Harry play a zero-sum game, represented by the pay-off matrix for Doug shown above. Use dominance to reduce the game to a 2 × 2 game. You must make your reasoning clear.

3

	Nick plays 1	Nick plays 2	Nick plays 3
Chris plays 1	1	2	3
Chris plays 2	−1	−3	1
Chris plays 3	2	−1	5

Chris and Nick play a zero-sum game, represented by the pay-off matrix for Chris shown above. Use dominance to reduce the game to a 2 × 2 game. You must make your reasoning clear.

Questions 4 to 7
a Verify that there is no stable solution.
b Determine the optimal mixed strategy and the value of the game to A.
c Determine the optimal mixed strategy and the value of the game to B.

4

	B plays 1	B plays 2
A plays 1	2	−4
A plays 2	−1	3

5

	B plays 1	B plays 2
A plays 1	−3	5
A plays 2	2	−4

6

	B plays 1	B plays 2
A plays 1	5	−1
A plays 2	−2	1

7

	B plays 1	B plays 2
A plays 1	−1	3
A plays 2	1	−2

5.8 You should be able to determine the optimal mixed strategy for the player with two choices in a 2 × 3 or 3 × 2 game.

When players have more than 2 choices you have a harder problem. (If A has three or more choices and you assign p as the probability that A plays row 1, you now cannot say that A will play row 2 with a probability of $1 - p$ since you have row 3 to add to your probability sum.)

If one player has only 2 choices it is always possible to determine their best strategy, so in a 2 × 3 game, you can determine A's mixed strategy, and in a 3 × 2 game you can determine B's mixed strategy. It is the player with 3 choices that causes the problem!

■ If there is more than one intersection point the probability giving the highest minimum point gives the solution.

Example 9

Alf and Bert play a zero-sum game, represented by the pay-off matrix below.

	Bert plays 1	Bert plays 2	Bert plays 3
Alf plays 1	−5	2	−1
Alf plays 2	2	−3	1

Determine Alf's best strategy and the value of the game to him.

	Bert plays 1	Bert plays 2	Bert plays 3	Row minimum
Alf plays 1	−5	2	−1	−5
Alf plays 2	2	−3	1	−3
Column maximum	2	2	1	

First check for a stable solution,

The row maximin is −3 and the column minimax is 1

−3 ≠ 1 so there is no stable solution.

Alf has only two choices so it will be possible to determine his strategy.

Let Alf play row 1 with probability p and row 2 with probability $1 − p$.

If Bert plays column 1   Alf's expected winnings are $V(A) = −5p + 2(1 − p) = 2 − 7p$

If Bert plays column 2   Alf's expected winnings are $V(A) = 2p − 3(1 − p) = 5p − 3$

If Bert plays column 3   Alf's expected winnings are $V(A) = −p + (1 − p) = 1 − 2p$

If you illustrate these on a diagram you get

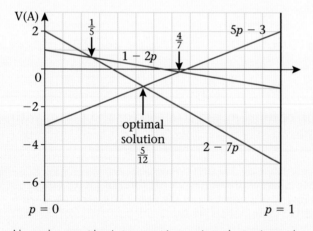

> Look from the 'bottom up' to find the highest point.

> This is Alf's minimum winnings. Choose the point that gives the greatest value for the minimum expected winnings. If Bert did not play sensibly Alf could win more than this. This is the least that Alf expects to win.

Now choose the intersection point that gives the highest minimum winnings.

This is where $2 − 7p = 5p − 3$, giving $p = \frac{5}{12}$, and $V(A) = −\frac{11}{12}$.

So Alf should play column 1 with a probability of $\frac{5}{12}$, and column 2 with a probability of $\frac{7}{12}$.

Example 10

A and B play a zero-sum game, given by the pay-off matrix below.

	B plays 1	B plays 2
A plays 1	4	−1
A plays 2	3	4
A plays 3	1	7

Determine B's best strategy, and the value of the game to B.

	B plays 1	B plays 2	Row minimum
A plays 1	4	−1	−1
A plays 2	3	4	3
A plays 3	1	7	1
Column maximum	4	7	

First check for a stable solution.

The row maximin is 3 and the column minimax is 4

$3 \neq 4$ so there is no stable solution

Player B has only two choices so it will be possible to determine his strategy.

Let B play column 1 with probability q and column 2 with probability $1 - q$.

If A plays row 1   B's expected winnings are $V(B) = -[4q - (1 - q)] = 1 - 5q$

If A plays row 2   B's expected winnings are $V(B) = -[3q + 4(1 - q)] = q - 4$

If A plays row 3   B's expected winnings are $V(B) = -[q + 7(1 - q)] = 6q - 7$

If we illustrate these on a diagram we get

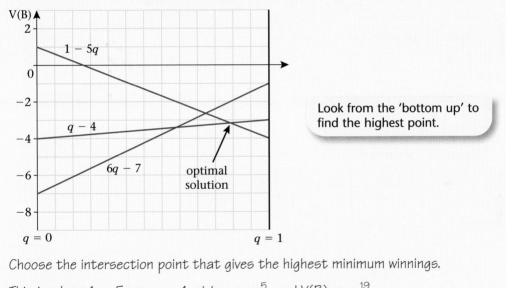

Look from the 'bottom up' to find the highest point.

optimal solution

Choose the intersection point that gives the highest minimum winnings.

This is where $1 - 5q = q - 4$, giving $q = \frac{5}{6}$, and $V(B) = -\frac{19}{6}$

So B should play column 1 with a probability of $\frac{5}{6}$, and column 2 with a probability of $\frac{1}{6}$.

Exercise 5C

For questions 1 to 4

a Verify that there is no stable solution.

b Determine the optimal mixed strategy and the value of the game to A.

1

	B plays 1	B plays 2	B plays 3
A plays 1	−5	2	2
A plays 2	1	−3	−4

2

	B plays 1	B plays 2	B plays 3
A plays 1	2	6	−2
A plays 2	−1	−4	3

3

	B plays 1	B plays 2	B plays 3
A plays 1	−2	3	6
A plays 2	5	1	−4

4

	B plays 1	B plays 2	B plays 3
A plays 1	5	−2	−4
A plays 2	−3	1	6

For questions 5 to 8

a Verify that there is no stable solution.

b Determine the optimal mixed strategy and the value of the game to B.

5

	B plays 1	B plays 2
A plays 1	−1	1
A plays 2	3	−4
A plays 3	−2	2

6

	B plays 1	B plays 2
A plays 1	−5	4
A plays 2	3	−3
A plays 3	1	−2

7

	B plays 1	B plays 2
A plays 1	−3	2
A plays 2	−1	−2
A plays 3	2	−4

8

	B plays 1	B plays 2
A plays 1	2	−3
A plays 2	−2	4
A plays 3	1	−1

5.9 You should be able to determine the optimal mixed strategy for the player with three choices in a 2 × 3 or 3 × 2 game.

- For a player with three choices there are two techniques you can use:
  - You can reduce the pay-off matrix using dominance arguments (see section 5.6)
  - If you cannot reduce the player's options to 2, use linear programming (see section 5.10)

5.10 You should be able to convert 2 × 3, 3 × 2 and 3 × 3 games into linear programming problems.

You use straight line graphs to represent a player's expected winnings and examine the intersection points to find the probability and value of the game. This is very similar to point testing when using graphical linear programming. If you have more than two variables, you use simplex.

In the examination you will have at most a 3 × 2, 2 × 3 or 3 × 3 game to solve.

Example 11

	B plays 1	B plays 2	B plays 3
A plays 1	−2	4	2
A plays 2	0	−3	−2
A plays 3	−6	1	3

a Formulate the game above as a linear programming problem for player A, writing the constraints as equalities and defining your variables clearly.

b Write down an initial simplex tableau for this problem.

c Outline how you would change the method to formulate the game as a linear programming problem for player B. (You do not need to formulate the problem)

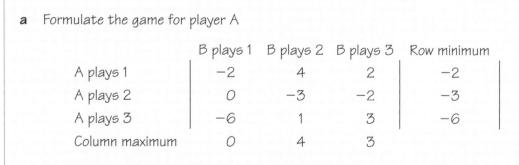

a Formulate the game for player A

	B plays 1	B plays 2	B plays 3	Row minimum
A plays 1	−2	4	2	−2
A plays 2	0	−3	−2	−3
A plays 3	−6	1	3	−6
Column maximum	0	4	3	

The row maximin is −2, the column minimax is 0

Since $-2 \neq 0$, the game is not stable.

You cannot use row or column domination to remove any rows or columns.

> You need to use simplex. You must make sure that the feasible region is completely in the positive 'quadrant'.

Adding 7 to each element gives

	B plays 1	B plays 2	B plays 3
A plays 1	5	11	9
A plays 2	7	4	5
A plays 3	1	8	10

> FIRST transform the game by adding n to each element, to make the values positive (>0).
> In this case you need to add 7 to every element.

Let p_1 be the probability of A playing row 1
Let p_2 be the probability of A playing row 2
Let p_3 be the probability of A playing row 3
Where $p_1, p_2, p_3 \geqslant 0$

> THEN define your decision variables.

Let $v =$ value of the original game to player A
Then $V = v + 7 =$ value of the new game to player A

Maximise $P = V$ so $P - V = 0$

> NEXT write down the objective function.

Subject to

If B plays column 1, $\quad 5p_1 + 7p_2 + p_3 \geqslant V$
If B plays column 2, $\quad 11p_1 + 4p_2 + 8p_3 \geqslant V$
If B plays column 3, $\quad 9p_1 + 5p_2 + 10p_3 \geqslant V$
$$p_1 + p_2 + p_3 = 1$$

> V is the minimum that A will win, so his expected winnings are $\geqslant V$.

> FINALLY write down the constraints.

Where $p_1, p_2, p_3 \geqslant 0$

so

$$V - 5p_1 - 7p_2 - p_3 \leqslant 0$$
$$V - 11p_1 - 4p_2 - 8p_3 \leqslant 0$$
$$V - 9p_1 - 5p_2 - 10p_3 \leqslant 0$$
$$p_1 + p_2 + p_3 \leqslant 1$$

> Notice that we have replaced $p_1 + p_2 + p_3 = 1$, by the weaker constraint $p_1 + p_2 + p_3 \leqslant 1$

b The linear programme is

Maximise $\textbf{P} - \textbf{V} = 0$

Subject to

$$V - 5p_1 - 7p_2 - p_3 + r = 0$$
$$V - 11p_1 - 4p_2 - 8p_3 + s = 0$$
$$V - 9p_1 - 5p_2 - 10p_3 + t = 0$$
$$p_1 + p_2 + p_3 + u = 1$$
$$p_1, p_2, p_3, r, s, t, u \geqslant 0$$

The initial simplex tableau is:

Basic variable	V	p_1	p_2	p_3	r	s	t	u	Value
r	1	−5	−7	−1	1	0	0	0	0
s	1	−11	−4	−8	0	1	0	0	0
t	1	−9	−5	−10	0	0	1	0	0
u	0	1	1	1	0	0	0	1	1
P	−1	0	0	0	0	0	0	0	0

c To formulate the game for player B:
  1 Rewrite the matrix from player B's point of view.
   • Transpose the matrix so all the rows become columns, and all the columns become rows.
   • Change the signs so positives become negatives, and negatives positives.
   In this example the game becomes

	A plays 1	A plays 2	A plays 3
B plays 1	2	0	6
B plays 2	−4	3	−1
B plays 3	−2	2	−3

  2 Continue with the method as usual:
   • make all the entries positive
   • define the variables
   • write down the objective and constraints
   • create the initial tableau
   • continue to apply the simplex algorithm.

Example 12

Jenny and Merry play a zero-sum game, represented by the pay-off matrix for Jenny shown below.

	Merry plays 1	Merry plays 2	Merry plays 3
Jenny plays 1	−3	0	2
Jenny plays 2	−2	2	1
Jenny plays 3	0	−3	−2

a Convert the game into a linear programming problem for Jenny, defining your decision variables. You should write the constraints as equations.

b Write down an initial simplex tableau making your variables clear.

c Solve the problem to determine Jenny's best strategy and the value of the game to her.

d Write down the initial simplex tableau you would use to determine Merry's best strategy.

a

	Merry plays 1	Merry plays 2	Merry plays 3	Row minimum
Jenny plays 1	−3	0	2	−3
Jenny plays 2	−2	2	1	−2
Jenny plays 3	0	−3	−2	−3
Column maximum	0	2	2	

The row maximin is −2, the column minimax is 0

$-2 \neq 0$ so there is no stable solution.

You cannot use row or column domination to remove any rows or columns.

In this case, add 4 to every element.

> FIRST transform the game by adding n to each element, to make the values positive (> 0).

So the game becomes

	M plays 1	M plays 2	M plays 3
J plays 1	1	4	6
J plays 2	2	6	5
J plays 3	4	1	2

> At the end you MUST remember to subtract this constant from the number that the tableau gives to find the value of the game.

Let Jenny play row 1 with probability p_1, row 2 with probability p_2 and row 3 with probability p_3.

$$p_1 + p_2 + p_3 = 1$$

If Merry plays column 1, Jenny's expected winnings are $p_1 + 2p_2 + 4p_3$

If Merry plays column 2, Jenny's expected winnings are $4p_1 + 6p_2 + p_3$

If Merry plays column 3, Jenny's expected winnings are $6p_1 + 5p_2 + 2p_3$

Now if you let V be the value of the game, Jenny will be looking for probabilities such that

$$p_1 + 2p_2 + 4p_3 \geq V$$
$$4p_1 + 6p_2 + p_3 \geq V$$
$$6p_1 + 5p_2 + 2p_3 \geq V$$

Rearrange these inequalities and introduce slack variables r, s and t

maximise

$$P = V$$

Subject to

$$V - p_1 - 2p_2 - 4p_3 + r = 0$$
$$V - 4p_1 - 6p_2 - p_3 + s = 0$$
$$V - 6p_1 - 5p_2 - 2p_3 + t = 0$$
$$p_1 + p_2 + p_3 + u = 1$$
$$p_1, p_2, p_3, r, s, t, u \geq 0$$

> u is really just a check variable. At the end the final tableau must give $u = 0$. The probabilities must total 1. In maximising V you will get $p_1 + p_2 + p_3 = 1$

b The initial tableau looks like this.

Basic variable	V	p_1	p_2	p_3	r	s	t	u	Value
r	1	−1	−2	−4	1	0	0	0	0
s	1	−4	−6	−1	0	1	0	0	0
t	1	−6	−5	−2	0	0	1	0	0
u	0	1	1	1	0	0	0	1	1
P	−1	0	0	0	0	0	0	0	0

c Solve this simplex problem in the usual way.

Basic variable	V	p_1	p_2	p_3	r	s	t	u	Value	θ values
r	1	−1	−2	−4	1	0	0	0	0	$0 \div 1 = 0$
s	1	−4	−6	−1	0	1	0	0	0	$0 \div 1 = 0$
t	1	−6	−5	−2	0	0	1	0	0	$0 \div 1 = 0$
u	0	1	1	1	0	0	0	1	1	$1 \div 0 = \infty$
P	−1	0	0	0	0	0	0	0	0	

> The first three rows all give the same θ values, they always will. It is useful to select the first row as the pivot row.

Basic variable	V	p_1	p_2	p_3	r	s	t	u	Value	Row operations
V	1	−1	−2	−4	1	0	0	0	0	R1 ÷ 1
s	0	−3	−4	3	−1	1	0	0	0	R2 − R1
t	0	−5	−3	2	−1	0	1	0	0	R3 − R1
u	0	1	1	1	0	0	0	1	1	No change
P	0	−1	−2	−4	1	0	0	0	0	R5 + R1

> Notice that, from this tableau, the top row (the first pivot row) and the bottom row (objective row) will be identical apart from the very first column. This is a useful time-saver.

Basic variable	V	p_1	p_2	p_3	r	s	t	u	Value	θ values
V	1	−1	−2	−4	1	0	0	0	0	
s	0	−3	−4	3	−1	1	0	0	0	$0 \div 3 = 0$
t	0	−5	−3	2	−1	0	1	0	0	$0 \div 2 = 0$
u	0	1	1	1	0	0	0	1	1	$1 \div 1 = 1$
P	0	−1	−2	−4	1	0	0	0	0	

> Remember that you can never choose a negative number as a pivot, so you do not need to consider the first row since you can not have −4 as a pivot.

> We could choose row 2 or row 3, since both have a zero θ value. Arbitrarily choose row 2 as the next pivot row.

Basic variable	V	p_1	p_2	p_3	r	s	t	u	Value	Row operations
V	1	-5	$-\frac{22}{3}$	0	$-\frac{1}{3}$	$\frac{4}{3}$	0	0	0	$R1 + 4\frac{R2}{3}$
p_3	0	-1	$-\frac{4}{3}$	1	$-\frac{1}{3}$	$\frac{1}{3}$	0	0	0	$R2 \div 3$
t	0	-3	$-\frac{1}{3}$	0	$-\frac{1}{3}$	$-\frac{2}{3}$	1	0	0	$R3 - 2\frac{R2}{3}$
u	0	2	$\frac{7}{3}$	0	$\frac{1}{3}$	$-\frac{1}{3}$	0	1	1	$R4 - \frac{R2}{3}$
P	0	-5	$-\frac{22}{3}$	0	$-\frac{1}{3}$	$\frac{4}{3}$	0	0	0	$R5 + 4\frac{R2}{3}$

The most negative number in the bottom row is -5, so the next pivot row must be row 4, since all other elements in the p_1 column are negative.

Basic variable	V	p_1	p_2	p_3	r	s	t	u	Value	Row operations
V	1	0	$-\frac{3}{2}$	0	$\frac{1}{2}$	$\frac{1}{2}$	0	$\frac{5}{2}$	$\frac{5}{2}$	$R1 + 5R4$
p_3	0	0	$-\frac{1}{6}$	1	$-\frac{1}{6}$	$\frac{1}{6}$	0	$\frac{1}{2}$	$\frac{1}{2}$	$R2 + R4$
t	0	0	$\frac{19}{6}$	0	$\frac{1}{6}$	$-\frac{7}{6}$	1	$\frac{3}{2}$	$\frac{3}{2}$	$R3 + 3R4$
p_1	0	1	$\frac{7}{6}$	0	$\frac{1}{6}$	$-\frac{1}{6}$	0	$\frac{1}{2}$	$\frac{1}{2}$	$R4 \div 2$
P	0	0	$-\frac{3}{2}$	0	$\frac{1}{2}$	$\frac{1}{2}$	0	$\frac{5}{2}$	$\frac{5}{2}$	$R5 + 5R4$

Basic variable	V	p_1	p_2	p_3	r	s	t	u	Value	Row operations
V	1	0	$-\frac{3}{2}$	0	$\frac{1}{2}$	$\frac{1}{2}$	0	$\frac{5}{2}$	$\frac{5}{2}$	
p_3	0	0	$-\frac{1}{6}$	1	$-\frac{1}{6}$	$\frac{1}{6}$	0	$\frac{1}{2}$	$\frac{1}{2}$	
t	0	0	$\frac{19}{6}$	0	$\frac{1}{6}$	$-\frac{7}{6}$	1	$\frac{3}{2}$	$\frac{3}{2}$	$\frac{3}{2} \div \frac{19}{6} = \frac{9}{19}$
p_1	0	1	$\frac{7}{6}$	0	$\frac{1}{6}$	$-\frac{1}{6}$	0	$\frac{1}{2}$	$\frac{1}{2}$	$\frac{1}{2} \div \frac{7}{6} = \frac{3}{7}$
P	0	0	$-\frac{3}{2}$	0	$\frac{1}{2}$	$\frac{1}{2}$	0	$\frac{5}{2}$	$\frac{5}{2}$	

The p_2 column is the next pivot column.

Once again the 4th row will become the pivot row.

Basic variable	V	p_1	p_2	p_3	r	s	t	u	Value	Row operations
V	1	$\frac{9}{7}$	0	0	$\frac{5}{7}$	$\frac{2}{7}$	0	$\frac{22}{7}$	$\frac{22}{7}$	$R1 + \frac{3}{2}\frac{R4}{2}$
p_3	0	$\frac{1}{7}$	0	1	$-\frac{1}{7}$	$\frac{1}{7}$	0	$\frac{4}{7}$	$\frac{4}{7}$	$R2 + \frac{1}{6}\frac{R4}{2}$
t	0	$-\frac{19}{7}$	0	0	$-\frac{2}{7}$	$-\frac{5}{7}$	1	$\frac{1}{7}$	$\frac{1}{7}$	$R3 - \frac{19}{6}\frac{R4}{2}$
p_2	0	$\frac{6}{7}$	1	0	$\frac{1}{7}$	$-\frac{1}{7}$	0	$\frac{3}{7}$	$\frac{3}{7}$	$R4 \div \frac{7}{6} = R4$
P	0	$\frac{9}{7}$	0	0	$\frac{5}{7}$	$\frac{2}{7}$	0	$\frac{22}{7}$	$\frac{22}{7}$	$R5 + \frac{3}{2}\frac{R4}{2}$

This tableau is optimal since there are no negatives in the objective row.

So Jenny should play   row 1 with a probability of 0

                       row 2 with a probability of $\frac{3}{7}$

                       row 3 with a probability of $\frac{4}{7}$

The value of the game to Jenny is $\frac{22}{7} - 4 = -\frac{6}{7}$

> u is not listed as a basic variable, so its value is 0, as required.

> Subtract 4 because 4 was added to each term at the start of **b**.

d • First re-write the game

	Jenny plays 1	Jenny plays 2	Jenny plays 3
Merry plays 1	3	2	0
Merry plays 2	0	−2	3
Merry plays 3	−2	−1	2

• Add 3 to each element

	Jenny plays 1	Jenny plays 2	Jenny plays 3
Merry plays 1	6	5	3
Merry plays 2	3	1	6
Merry plays 3	1	2	5

• Define the decision variables.

Let Merry play row 1 with probability q_1, row 2 with probability q_2 and row 3 with probability q_3.
Let the value of the game to Merry be V.

• Formulate the problem.

Maximise $P = V$
Subject to

$$V - 6q_1 - 3q_2 - q_3 + r = 0$$
$$V - 5q_1 - q_2 - 2q_3 + s = 0$$
$$V - 3q_1 - 6q_2 - 5q_3 + t = 0$$
$$q_1 + q_2 + q_3 + u = 1$$
$$p_1, p_2, p_3, r, s, t, u \geqslant 0$$

• The initial tableau is

Basic variable	V	q_1	q_2	q_3	r	s	t	u	Value
r	1	−6	−3	−1	1	0	0	0	0
s	1	−5	−1	−2	0	1	0	0	0
t	1	−3	−6	−5	0	0	1	0	0
u	0	1	1	1	0	0	0	1	1
P	−1	0	0	0	0	0	0	0	0

Exercise 5D

Questions 1 to 4

Formulate the games below as linear programming problems for player A, writing the constraints as equalities and clearly defining your variables.

1

	B plays 1	B plays 2
A plays 1	-1	1
A plays 2	3	-4
A plays 3	-2	2

2

	B plays 1	B plays 2	B plays 3
A plays 1	-5	4	1
A plays 2	3	-3	2
A plays 3	1	-2	-1

3

	B plays 1	B plays 2	B plays 3
A plays 1	-3	2	-1
A plays 2	-1	-2	1
A plays 3	2	-4	-2

4

	B plays 1	B plays 2	B plays 3
A plays 1	2	-3	-1
A plays 2	-2	4	1
A plays 3	1	-1	0

5 Formulate the game below as a linear programming problem for player B, writing the constraints as equalities and clearly defining your variables.

	B plays 1	B plays 2	B plays 3
A plays 1	-5	2	3
A plays 2	1	-3	-4

Questions 6 to 8

Formulate the games below as linear programming problems for player B, writing the constraints as equalities and clearly defining your variables.

6

	B plays 1	B plays 2	B plays 3
A plays 1	-5	4	1
A plays 2	3	-3	2
A plays 3	1	-2	-1

7

	B plays 1	B plays 2	B plays 3
A plays 1	−3	2	−1
A plays 2	−1	−2	1
A plays 3	2	−4	−2

8

	B plays 1	B plays 2	B plays 3
A plays 1	2	−3	−1
A plays 2	−2	4	1
A plays 3	1	−1	0

9 Using your answer to question **1**,

  a write down an initial simplex tableau to solve the zero-sum game below, for player A,

  b use the simplex algorithm to determine A's best strategy.

	B plays 1	B plays 2
A plays 1	−1	1
A plays 2	3	−4
A plays 3	−2	2

10 Using your answer to question **5**,

  a write down an initial simplex tableau to solve the zero-sum game below, for player B,

  b use the simplex algorithm to determine B's best strategy.

	B plays 1	B plays 2	B plays 3
A plays 1	−5	2	3
A plays 2	1	−3	−4

11

	B plays 1	B plays 2	B plays 3
A plays 1	−5	4	1
A plays 2	3	−3	2
A plays 3	1	−2	−1

Using your answer to question **2**,

a write down an initial simplex tableau to solve the zero-sum game, for player A,

b use the simplex algorithm to determine A's best strategy.

Using your answer to question **6**,

c write down an initial simplex tableau to solve the zero-sum game, for player B,

d use the simplex algorithm to determine B's best strategy.

12

	B plays 1	B plays 2	B plays 3
A plays 1	−3	2	−1
A plays 2	−1	−2	1
A plays 3	2	−4	−2

Using your answer to question **3**,

a write down an initial simplex tableau to solve the zero-sum game, for player A,

b use the simplex algorithm to determine A's best strategy.

Using your answer to question **7**,

c write down an initial simplex tableau to solve the zero-sum game, for player B,

d use the simplex algorithm to determine B's best strategy.

Mixed exercise 5E

1 A two-person zero-sum game is represented by the following pay-off matrix for player A. Find the best strategy for each player and the value of the game.

		B	
		I	II
A	I	4	−2
	II	−5	6

2 Ben and Greg play a zero-sum game, represented by the following pay-off matrix for Ben.

	Greg plays 1	Greg plays 2	Greg plays 3
Ben plays 1	−5	4	3
Ben plays 2	1	−1	−4

a Explain why this matrix might be reduced to

$$\begin{vmatrix} -5 & 3 \\ 1 & -4 \end{vmatrix}$$

b Hence find the best strategy for each player and the value of the game.

3 Cait and Georgi play a zero-sum game, represented by the following pay-off matrix for Cait.

	Georgi plays 1	Georgi plays 2	Georgi plays 3
Cait plays 1	−5	2	3
Cait plays 2	1	−3	−4
Cait plays 3	−7	0	1

a Identify the play safe strategies for each player.

b Verify that there is no stable solution to this game.

c Use dominance to reduce the game to a 2 × 3 game, explaining your reasoning.

d Find Cait's best strategy and the value of the game to her.

e Write down the value of the game to Georgi.

4 A two person zero-sum game is represented by the following pay-off matrix for player A.

	B plays 1	B plays 2	B plays 3
A plays 1	2	−1	−3
A plays 2	−2	1	4
A plays 3	−3	1	−3
A plays 4	−1	2	−2

a Verify that there is no stable solution to this game

b Explain the circumstances under which a row, x, dominates a row, y.

c Reduce the game to a 3×3 game, explaining your reasoning.

d Formulate the 3×3 game as a linear programming problem for player A. Write the constraints as inequalities and define your variables.

5 A two person zero-sum game is represented by the following pay-off matrix for player A.

	B plays 1	B plays 2	B plays 3
A plays 1	5	−3	1
A plays 2	−1	−4	4
A plays 3	3	2	−1

a Identify the play safe strategies for each player.

b Verify that there is no stable solution to this game.

c Use dominance to reduce the game to a 3×2 game, explaining your reasoning.

d Write down the pay-off matrix for player B.

e Find B's best strategy and the value of the game.

6 A two person zero-sum game is represented by the following pay-off matrix for player A.

	B plays 1	B plays 2	B plays 3
A plays 1	2	7	−1
A plays 2	5	0	8
A plays 3	−2	3	5

a Identify the play safe strategies for each player.

b Verify that there is no stable solution to this game.

c Write down the pay-off matrix for player B

d Formulate the game for player B as a linear programming problem. Define your variables and write your constraints as equations.

e Write down an initial tableau that you could use to solve the game for player B.

Summary of key points

1 A two person game is one in which only two parties can play.

2 When playing safe each player looks for the worst that could happen if s/he makes each choice in turn. S/he then picks the choice that results in the least worst option.

3 A zero-sum game is one in which the two entries in each cell add up to zero.

4 A pay-off matrix is always written from the row player's point of view unless you are told otherwise.

5 The play safe strategies are
- the row maximin for the row player
- the column minimax for the column player.

6 In a zero-sum game there will be a stable solution if and only if

the row maximin = the column minimax

7 If one named row (or column) is *always* a better option than another named row (or column) then the better option **dominates**.

8 The pay-off matrix can be reduced by deleting a dominated row or column.

9 To determine an optimal strategy, form probability equations and graph them.

10 The intersection point gives you the information you need to work out the probabilities and the value of the game.

11 If there is more than one intersection point the probability giving the highest minimum point gives the solution.

12 For a player with three choices:
- use dominance to reduce the pay-off matrix
- if you cannot do this, use linear programming.

13 To use simplex:
- transform the game by adding n to each element to make all values > 0
- define the decision variables
- write down the objective function
- write down the constraints.

In this chapter you will:

- learn some of the terminology used in analysing flows through networks
- be able to find an initial flow through a capacitated directed network
- start from an initial flow and use the labelling procedure to find flow-augmenting routes to increase the flow through the network
- confirm that a flow is maximal using the maximum flow–minimum cut theorem
- adapt the algorithm to deal with networks with multiple sources and/or sinks.

Network flows

In this chapter you will be looking at flow through a directed network. This could be oil or water or gas through a network of pipes, traffic through a one-way system, passengers through an airport, people through a system of fire escape corridors, or money between banks. It has also been applied to airline scheduling, security of statistical data, open-pit mining and railway networking.

6.1 You should be familiar with some of the terminology used in analysing flows through networks.

■ This chapter deals with **capacitated directed networks**, which are also called **capacitated directed graphs**, or **capacitated digraphs**. In a directed network each arc has an arrow indicating the permitted direction of flow.

■ Each arc will have a weight, and each weight represents the **capacity** of that arc. The capacity is the maximum amount of flow that can pass along that arc.

■ A vertex, S, is called a **source** if all arcs containing S are directed away from S. A vertex, T, is called a **sink** if all arcs containing T are directed towards T.

> In network flow questions, S and T are commonly used to denote source and sink vertices. It is possible to have more than one source and/or more than one sink – see Section 6.6

Example 1

In the digraph

a write down the capacity of arcs DF and BE.

b State the source vertex.

c State the sink vertex.

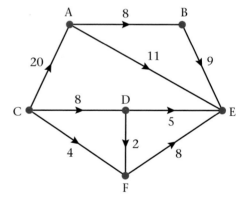

a DF has capacity 2 and BE has a capacity of 9. •————— This means that the maximum flow along DF is 2 – from D to F, and the maximum flow along BE is 9 – in the direction from B to E.

b The source vertex is C. •————— All the arcs that include C have arrows that are directed away from C.

c The sink vertex is E. •————— All the arcs that include E have arrows that are directed towards E.

- To show a **flow** through a network, assign a (non-negative) number to each arc so that it satisfies two conditions.
  - The **feasibility condition** which says that the flow along each arc must not exceed the capacity of that arc.
  - The **conservation condition** on all but source and sink vertices, which says that

    the total flow into a vertex = the total flow out of the vertex

  so the flow cannot 'build up' at a vertex.

- If the arc contains a flow equal to its capacity we say that arc is **saturated**.

Example **2**

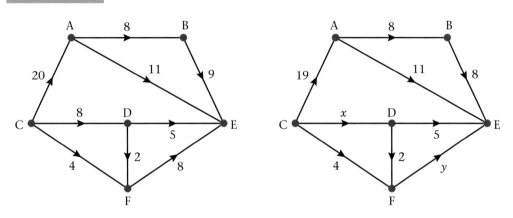

The diagram shows the directed network from example 1 and a possible flow through the network.

a Use flow conservation to find the values of x and y, explain your reasoning.

b List the five saturated arcs.

c Write down the value of the flow.

d What is the current flow along route CAE?

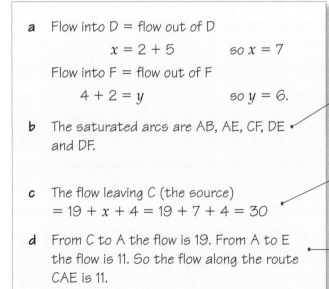

a Flow into D = flow out of D

$$x = 2 + 5 \qquad \text{so } x = 7$$

Flow into F = flow out of F

$$4 + 2 = y \qquad \text{so } y = 6.$$

b The saturated arcs are AB, AE, CF, DE and DF.

c The flow leaving C (the source)
$$= 19 + x + 4 = 19 + 7 + 4 = 30$$

d From C to A the flow is 19. From A to E the flow is 11. So the flow along the route CAE is 11.

We are looking for arcs where the flow is equal to the capacity.

We could have calculated the flow by looking at the flow arriving at E (the sink)
$$8 + 11 + 5 + y = 8 + 11 + 5 + 6 = 30.$$

When asked to find the flow along a particular route you need to find the *maximum* value that can pass along each arc. In this case, although 19 can start along the given route, only 11 can continue along AE, so only 11 can travel along this route.

Often you draw one diagram showing the directed network, together with a feasible flow. Usually the flow is indicated by circled numbers.

Example 3

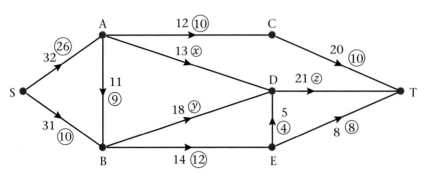

The diagram shows a capacitated directed network. The number on each arc represents the capacity of that arc. The numbers in circles represent an initial flow pattern.

a State a saturated arc.

b Find the values of x, y and z, explaining your reasoning.

c State the value of the initial flow.

d Write down
  i the capacity of arc BE
  ii the current flow along arc BE.

e Find the current flow along SABET.

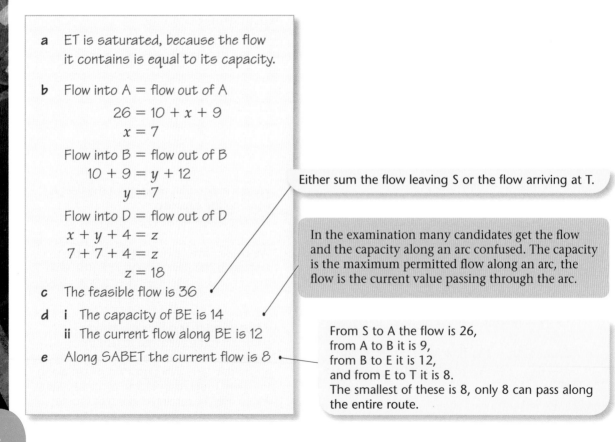

a ET is saturated, because the flow it contains is equal to its capacity.

b Flow into A = flow out of A
$$26 = 10 + x + 9$$
$$x = 7$$
Flow into B = flow out of B
$$10 + 9 = y + 12$$
$$y = 7$$
Flow into D = flow out of D
$$x + y + 4 = z$$
$$7 + 7 + 4 = z$$
$$z = 18$$

c The feasible flow is 36

d **i** The capacity of BE is 14
  ii The current flow along BE is 12

e Along SABET the current flow is 8

Either sum the flow leaving S or the flow arriving at T.

In the examination many candidates get the flow and the capacity along an arc confused. The capacity is the maximum permitted flow along an arc, the flow is the current value passing through the arc.

From S to A the flow is 26,
from A to B it is 9,
from B to E it is 12,
and from E to T it is 8.
The smallest of these is 8, only 8 can pass along the entire route.

Exercise 6A

There are photocopy masters on the CD-ROM for Questions 5 and 6

1

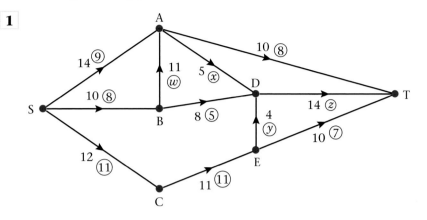

The diagram shows a capacitated directed network. The number on each arc represents the capacity of that arc. The numbers in circles represent an initial flow pattern.

a Find the values of w, x, y and z, explaining your reasoning.

b State the value of the initial flow.

c Identify two saturated arcs.

d Write down the capacity of arc BD.

e What is the current flow along route SAT?

2

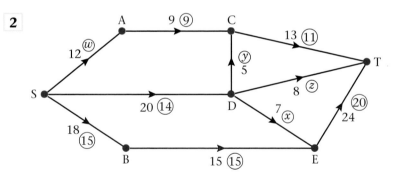

The diagram shows a capacitated directed network. The number on each arc represents the capacity of that arc. The numbers in circles represent an initial flow pattern.

a Find the values of w, x, y and z, explaining your reasoning.

b State the value of the initial flow.

c Identify two saturated arcs.

d Write down the flow along arc SD.

e What is the current flow along the route SBET?

3

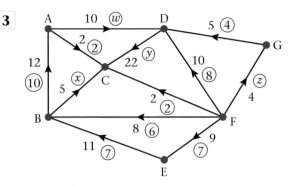

The diagram shows a capacitated directed network. The number on each arc represents the capacity of that arc. The numbers in circles represent an initial flow pattern.

a State the source vertex.

b State the sink vertex.

c Find the values of w, x, y and z, explaining your reasoning.

d State the value of the feasible flow.

e Identify three saturated arcs.

f Write down the capacity of arc FB.

4

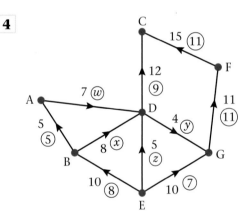

The diagram shows a capacitated directed network. The number on each arc represents the capacity of that arc. The numbers in circles represent an initial flow pattern.

a State the source vertex.

b State the sink vertex.

c Find the values of w, x, y and z, explaining your reasoning.

d State the value of the initial flow.

e Identify four saturated arcs.

f Write down the flow along arc FC.

5

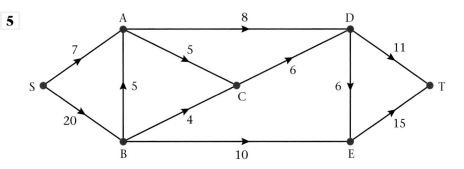

The diagram shows a capacitated directed network. The number on each arc represents the capacity of that arc. Find a feasible flow of at least 20 through the network from S to T.

6

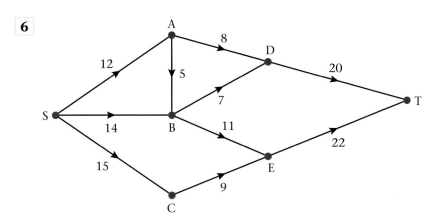

The diagram shows a capacitated directed network. The number on each arc represents the capacity of that arc. Find a feasible flow of 32 through the network from S to T.

6.2 You should understand what is meant by a cut.

■ A **cut**, in a network with source S and sink T, is a set of arcs whose removal separates the network into two parts X and Y, where X contains at least S and Y contains at least T.

> A cut divides the network into two. One piece **must** contain the source(s) and the other **must** contain the sink(s).

■ The **capacity (value) of a cut** is the sum of the capacities of those arcs in the cut which are directed from X to Y.

> Note that we use the **capacities** of the **arcs** to find the **capacity** of the **cut**. In the examination many candidates try to use the flows along the arcs instead of the capacities.

> You may be asked to write down these definitions in the examination.

Example 4

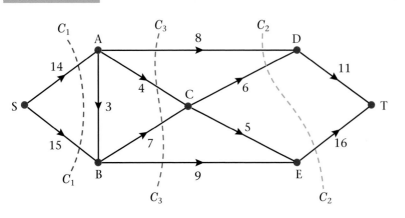

Determine the capacities of cuts C_1, C_2 and C_3.

Cut C_1 breaks the network into two pieces like this.

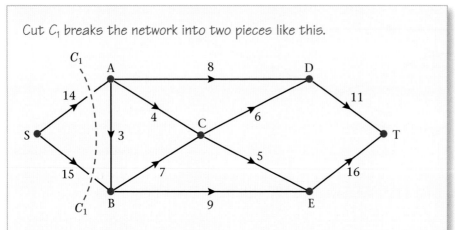

It creates two **vertex sets** {S} and {A, B, C, D, E, T}. These list which vertices are now in which 'broken half' of the network.

To evaluate the capacity of the cut it might help to imagine that it is a network of water pipes that have been broken, and the water is gushing out onto the floor, creating a puddle. Draw the puddle by connecting the top and bottom of the cut by drawing a line **behind** the sink.

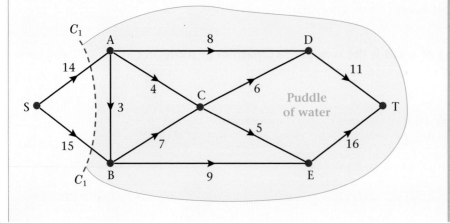

To find the capacity of the cut you need to see 'how much water is flowing onto the carpet'. We assume that the broken pipes are flowing at capacity. So you have 14 coming from SA and 15 from SB, giving a total of 29.

Cut $C_1 = 14 + 15 = 29$

'Drawing the puddle' for cut C_2 we get

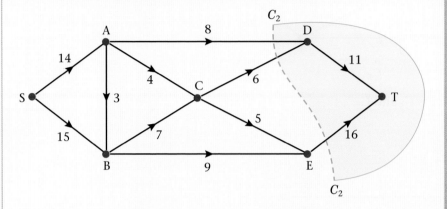

Flowing 'into the puddle' are AD value 8, CD value 6 and ET value 16. So the capacity of the cut is $8 + 6 + 16 = 30$

Cut $C_2 = 8 + 6 + 16 = 30$

'Drawing the puddle' for cut C_3 we get

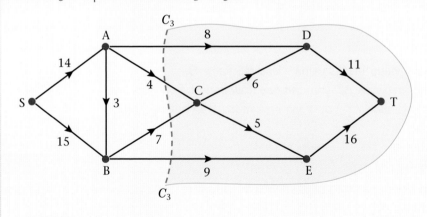

Flowing 'into the puddle' are AD value 8, AC value 4, BC value 7 and BE value 9.
So the capacity of the cut is $8 + 4 + 7 + 9 = 28$

Cut $C_3 = 8 + 4 + 7 + 9 = 28$

Arcs directed from the source cut set to the sink cut set are said to flow *into* the cut. Those directed from the sink to source sets are said to flow *out* of the cut.

- When evaluating the capacity of a cut, only include the capacities of the arcs flowing **into** the cut.

- Arcs which are cut, but whose **direction** flows **out** of the cut, contribute **zero** to the capacity of the cut

Example 5

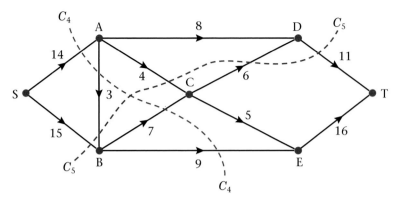

Determine the values of cuts C_4 and C_5.

'Drawing the puddle' for cut C_4 we get

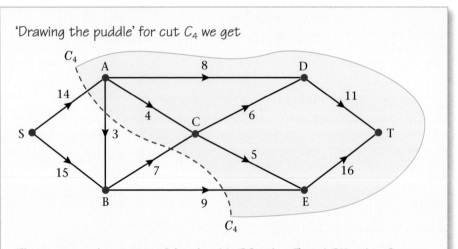

Flowing into the cut are SA value 14, BC value 7 and BE value 9.
If you look carefully at the arrow on AB you can see it does not flow
into the cut (puddle), so although AB has been broken it contributes
nothing to the value of the cut.
(In practical terms you can see that the supply to A has been
severed when you cut SA, so the 'pipe' AB will be empty.)
So for this cut, only sum the capacities of arcs SA, BC and BE.

Cut $C_4 = 14 + 7 + 9 = 30$

'Drawing the puddle' for cut C_5 we get

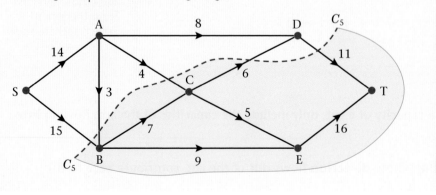

Flowing into the cut are DT value 11, AC value 4, AB value 3 and SB value 15.

CD, although cut, does not contribute to the value of the cut, since its direction indicates that it flows out of the cut.

Notice also that you did not include arc AB in cut C_4, but it has been included in cut C_5.

$$\text{Cut } C_5 = 11 + 4 + 3 + 15 = 33$$

You will have noticed that 5 cuts on the same network gave four different values, ranging from 28 to 33. In general different cuts will give different values.

■ **When a diagram is given that shows both capacities and a flow, the value of the cut is calculated using the capacities.**

Example 6

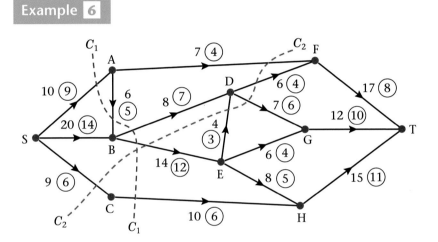

The diagram shows a capacitated directed network. The number on each arc represents the capacity of that arc. The numbers in circles represent an initial flow pattern.

Find the values of cuts C_1 and C_2.

Drawing cut C_1 we get

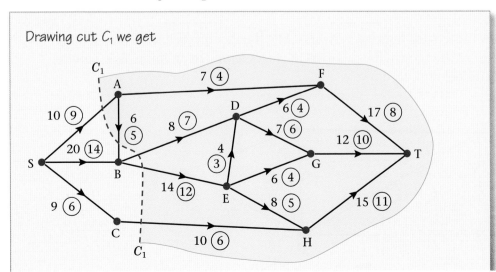

Flowing into the cut are arcs SA, BD, BE and CH.

AB flows out of the cut so contributes zero to its value.

Use the capacity numbers to evaluate the cut.

So the value of the cut = 10 (SA) + 8 (BD) + 14 (BE) + 10 (CH).

$C_1 = 10 + 8 + 14 + 10 = 42$

> Many candidates sum the flow values in the examination, which is incorrect.

Drawing cut C_2 we get

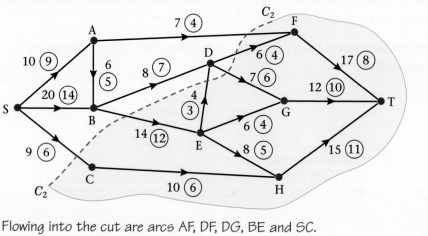

Flowing into the cut are arcs AF, DF, DG, BE and SC.

ED flows out of the cut and so contributes zero to its value.

Use the capacity numbers – not the flow values.

So the value of the cut = 7 (AF) + 6 (DF) + 7 (DG) + 14 (BE) + 9 (SC)

$C_2 = 7 + 6 + 7 + 14 + 9 = 43$

Exercise 6B

Questions 1 to 6

The diagrams show capacitated directed networks. The number on each arc represents the capacity of that arc. Where shown, the numbers in circles represent an initial flow pattern. Evaluate the capacities of the cuts drawn.

1

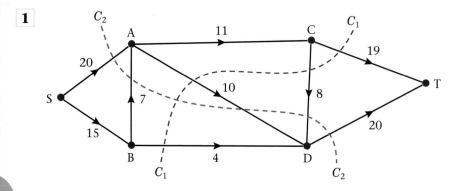

2

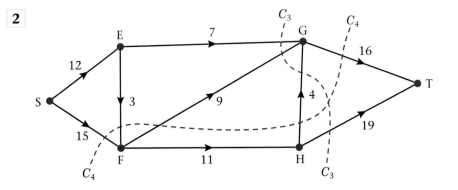

3

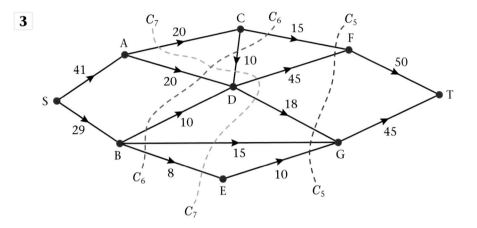

4

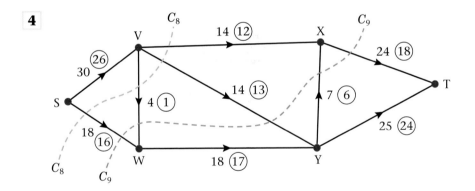

5

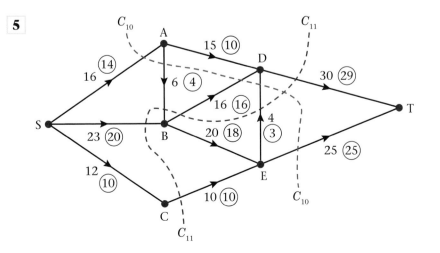

6

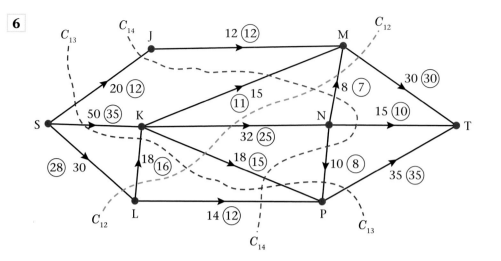

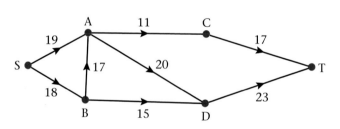

6.3 You should be able to find an initial flow through a capacitated directed network.

■ In the examination you will often be given an initial flow, but you may be directed to find it.

Example 7

S — A (19)
A — C (11)
A — B (17)
A — D (20)
B — D (15)
C — T (17)
D — T (23)

The diagram shows a capacitated directed network. The number on each arc represents the capacity of that arc.

a State the maximum flows along SACT and SBADT.

b Show these on a diagram.

Using this as your initial flow pattern,

c calculate the value of the initial flow.

---

a The maximum flow along SACT is 11.

> The arcs are SA (19), AC (11) and CT (17). The greatest flow along that route is 11. The flow is governed by the arc with the minimum capacity.

The maximum flow along SBADT is 17.

> The arcs are SB (18), BA (17), AD (20), DT (23). The greatest flow possible is 17. Arc BA has the lowest capacity and so this gives the maximum possible flow along this path.

b

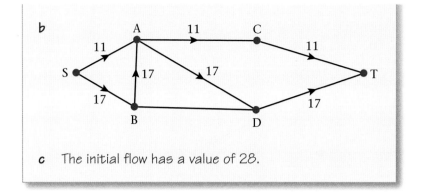

c The initial flow has a value of 28.

The total amount leaving S
(or arriving at T) is 28.

Example 8

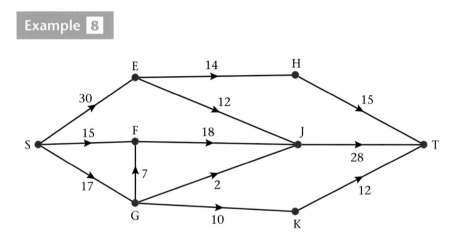

The diagram shows a capacitated directed network. The number on each arc represents the capacity of that arc.

Given that arcs SG, EH, EJ, GJ, GK and JT are saturated, draw an initial flow through the network.

Start by noting the saturated arcs. Saturated means that these six arcs are at capacity, so you can place these six numbers onto the diagram.

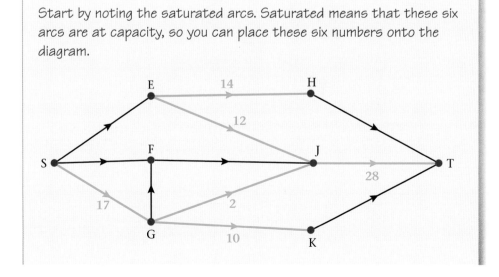

Now use flow-conservation to deduce some other values.
Look at the flow into H and into K. You can add two more numbers.

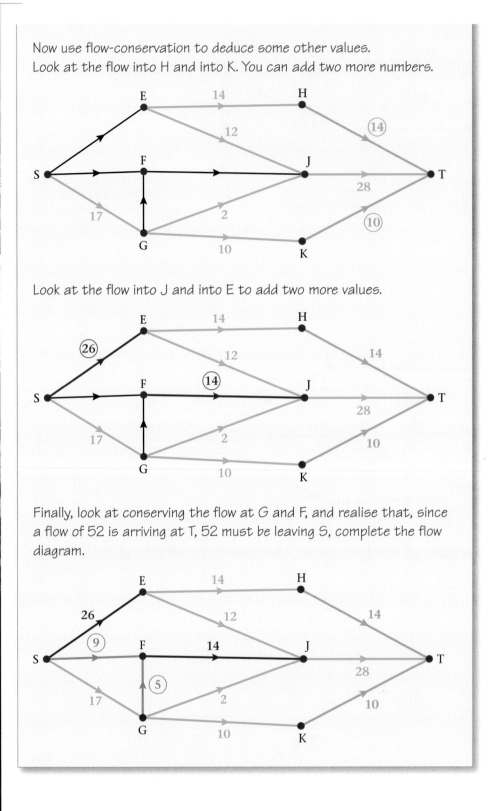

Look at the flow into J and into E to add two more values.

Finally, look at conserving the flow at G and F, and realise that, since a flow of 52 is arriving at T, 52 must be leaving S, complete the flow diagram.

Exercise 6C

There are photocopy masters on the CD-ROM for the questions in this exercise.

1

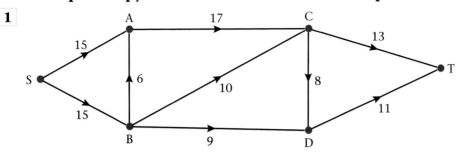

The diagram shows a capacitated directed network. The number on each arc represents the capacity of that arc.

a State the maximum flows along SACT and SBCDT.

b Show these on a diagram.

Using this as your initial flow,

c calculate the value of the initial flow.

2

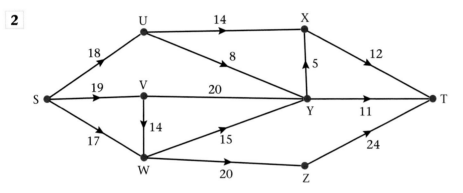

The diagram shows a capacitated directed network. The number on each arc represents the capacity of that arc.

a State the maximum flows along SUXT, SWZT and SVWYT.

b Show these on a diagram.

Using this as your initial flow,

c calculate the value of the initial flow.

3

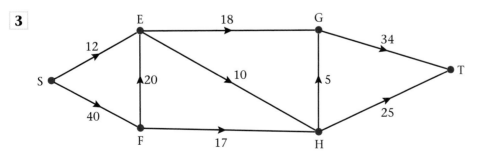

The diagram shows a capacitated directed network. The number on each arc represents the capacity of that arc.

a Given that arcs SE, EG, EH, FH and HG are saturated, draw an initial flow through the network.

b State the value of the initial flow.

4

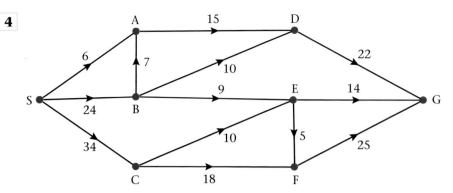

The diagram shows a capacitated directed network. The number on each arc represents the capacity of that arc.

a Given that arcs SB, BA, BD, CE, CF, EF and DG are saturated, draw an initial flow through the network.

b State the value of the initial flow.

6.4 **Starting from an initial flow you can use the labelling procedure to find flow-augmenting routes to increase the flow through the network.**

Although it is possible to find flows by inspection on small, simple diagrams, you need an algorithm that you can apply to more complicated networks that will guarantee finding the maximum possible flow through the network.

It may be necessary to re-route the initial flow, so you use the **labelling procedure** which allows you to reduce the flow along some arcs as well as to increase it in others. You use this to find **flow-augmenting** routes to increase the flow.

■ **In the labelling procedure, you draw two arrows on each arc.**

• **The 'forward' arrow (the arrow in the same direction as the arc) is used to identify the amount by which the flow along that arc can be increased. (This is the spare capacity.)**

• **The 'backward' arrow is used to identify the amount by which the flow in the arc could be reduced. (It shows the value of the current flow in that arc but not the direction.)**

• **Each flow-augmenting route starts at S and finishes at T.**

• **You may use both forward and backward arrows to find a route, but there must be capacity in the direction in which you wish to move.**

Example **9**

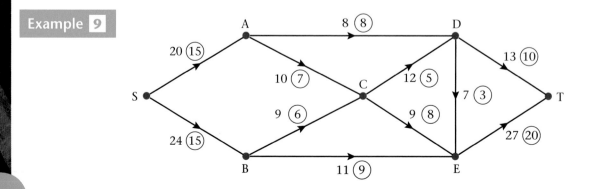

The diagram shows a capacitated directed network. The number on each arc represents the capacity of that arc. The numbers in circles represent an initial flow pattern.

Starting from this initial flow pattern,

a use the labelling procedure to obtain a maximum flow pattern through the network from S to T. You should list each flow-augmenting route you use together with its flow.

b Draw your final flow pattern.

a Apply the labelling procedure.

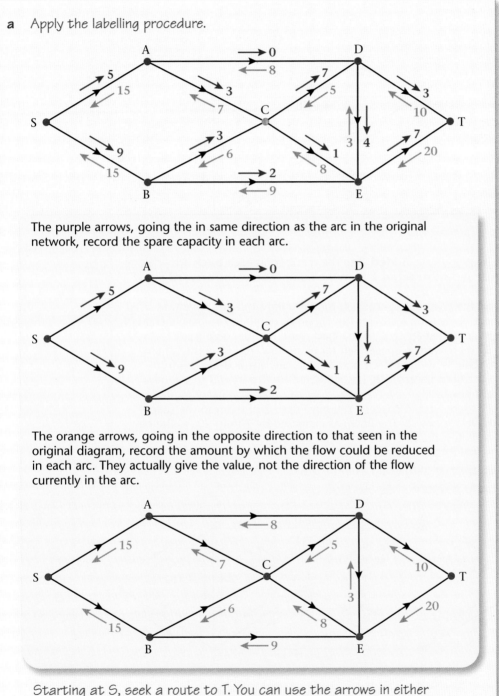

The purple arrows, going the in same direction as the arc in the original network, record the spare capacity in each arc.

The orange arrows, going in the opposite direction to that seen in the original diagram, record the amount by which the flow could be reduced in each arc. They actually give the value, not the direction of the flow currently in the arc.

Starting at S, seek a route to T. You can use the arrows in either direction, but there must be spare capacity in the direction you wish to move.

We often refer to the arrows that go in the direction we wish to move as **forward arrows**, and those going in the opposite direction **backward arrows.**

The sense of 'forward' and 'backward' is determined by the direction in we wish to move along the arc.

There are many different routes that could be chosen.

You could send an additional flow of 2 along route SBET (BE has maximum spare capacity of 2).
The updated diagram looks like this.

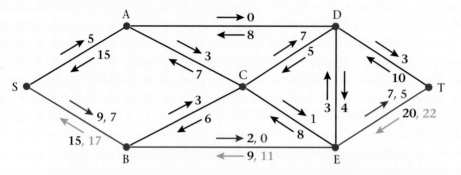

Notice that you have **added 2** to the number on each **back** arrow along the route and **subtracted 2** from the number on each **forward** arrow along the route.

Along SB there were 9 'spaces' left for extra flow, but you have now used 2 of them up, leaving just 7 for the 'forward' number. You have increased the flow along that arc by 2 so the 'backward' number goes from 15 to 17.

Along BE there were 2 'spaces' left for extra flow. You have used them both so there is no 'space' left and the forward arrow decreases from 2 to 0. The flow along arc BE has increased by two from 9 so the new backward flow is 11.

Along ET there were 7 'spaces'. You have used 2 of them so the spaces for extra flow is now 5. The flow has increased by 2 from 20, so the new 'backward' flow is 22.

You could send an additional flow of 3 along SACDT.
The updated diagram looks like this.

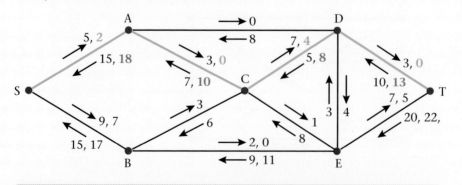

In the examination you do not need to draw each stage of the updated diagrams, you can just update your initial diagram.

You **subtract** three from each **forward** number along the route and **add** three to each **backward** number along the route.

So you subtract three from the numbers in the direction you move, and add three to the numbers on the arrows going against your direction of travel.

Now look for further routes.

There is no point in looking for routes beginning SA, because all arcs leaving A are now saturated – their 'forward' arrows are zero, indicating that there is no spare capacity.

If you start SB you then **must** go to C, since there is no other unsaturated exit from B. So your route must start SBC. From C you have a choice CD or CE. You may choose either. Arbitrarily you choose SBCD. You then **must** use DE – there is no other unsaturated exit available from D. This gives the route as SBCDET.

The maximum extra flow along SBCDET is 3.

The updated diagram looks like this.

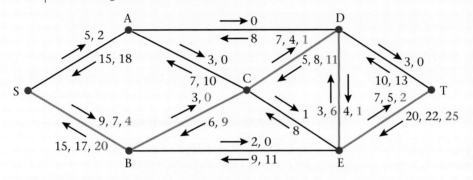

This is the last route. You cannot start SA. All arcs leaving A are now saturated, their forward numbers are zero.

You cannot start SB, since all the arcs leaving B have zero as their forward number, indicating that they too are saturated.

Thus there are no further flow-augmenting routes to find, and you have increased the flow through the network to its maximum.

b   Final flow pattern.

Start by making sure you have the arrows on each arc in the correct direction. Use the original diagram given in the question. Be particularly careful of the direction of any 'vertical' arrows – DE in this case.

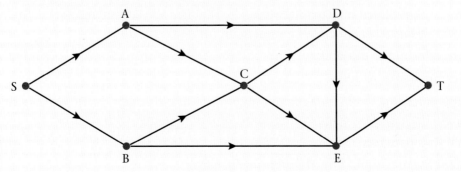

Then look at the 'backward' arrows, since they show the flow along each arc, and use the most recently updated numbers.

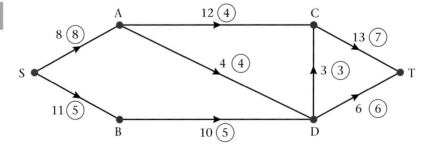

For example, look at arc CD. The arrow goes from C to D, so you are looking at the 5, 8, 11 list of numbers. Of these the last one written was 11, so the flow along CD is recorded as 11.

You should find that the flow into each intermediate vertex is equal to the flow leaving it, and the flow leaving S is equal to the flow arriving at T.

- When using the labelling procedure to find a flow-augmenting route, you may move in either direction along an arc, as long as there is not a zero in the direction you wish to move.

- When sending an extra flow of value f along a flow-augmenting route, you

  Subtract f from the number on each 'forward' arrow in the route.

  Add f to the number on each 'backward' arrow in the route.

- Sometimes it is necessary to make use of a backflow when finding a flow-augmenting route.

Example 10

The diagram shows a capacitated directed network. The number on each arc represents the capacity of that arc. The numbers in circles represent an initial flow pattern.

a Write down the value of the initial flow.

Starting from this initial flow pattern,

b use the labelling procedure to obtain a maximum flow pattern through the network from S to T. You should list each flow-augmenting route you use together with its flow.

c Draw your final flow pattern.

d State the value of the maximum flow through this network.

a The value of the initial flow is 13.

> Look at the numbers in circles. There is a flow of 5 + 8 leaving S and a flow of 6 + 7 arriving at T.

b Apply the labelling procedure:

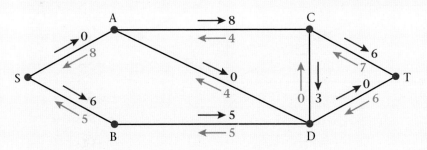

You can see that SA is saturated, because there is a zero in the direction S to A.

You must start SBD.

From D the arcs DC and DT have zero capacity in the direction you need to move, so your only exit is DA.

From A you can go to C and then onto T.

This gives your flow-augmenting route as SBDACT.

The smallest 'forward' number along this route is 4, giving the value of the flow.

> On the original diagram the direction of flow is from A to D and in this flow-augmenting route you are going from D to A. This is an example of a **backflow**.
> Many students are worried about backflows, but all you are doing is re-routing the initial flow pattern. You are reducing the flow along AD and re-directing it along AC.

The updated diagram is

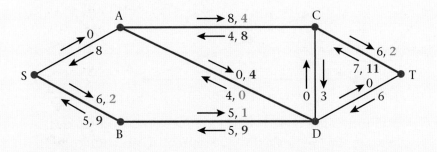

> Notice that in updating the numbers along AD, you have reduced the number going 'forwards', from D to A, by 4 since your flow-augmenting route went from D to A. You have similarly increased the number going 'backwards', from A to D, in our route by 4.

There are no further flow-augmenting routes you can use, so your flow is maximal.

c Start by putting all the arrows on the arcs, using the original diagram.

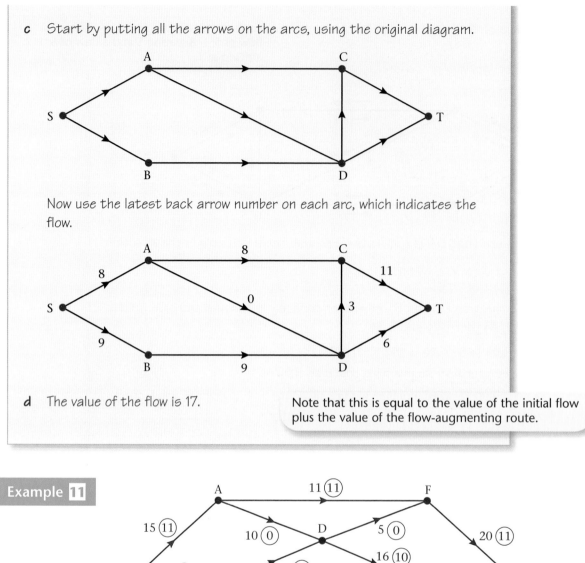

Now use the latest back arrow number on each arc, which indicates the flow.

d The value of the flow is 17.

> Note that this is equal to the value of the initial flow plus the value of the flow-augmenting route.

Example 11

The diagram shows a network of corridors represented by arcs. The corridors are those that will be used in evacuating a large cinema in the event of a fire. The capacity of the corridor is shown on each arc. The numbers in circles represent a possible flow of 41 people per second, from S to T.

In order for the building to achieve a safety certificate it is necessary to show that it is possible for at least 70 people per second to pass from S to T.

a Use the initial flow and the labelling procedure to find a maximum flow through the network. You must list each flow-augmenting route you use, together with its flow.

b Draw your maximal flow pattern.

c Will the cinema achieve its safety certificate?

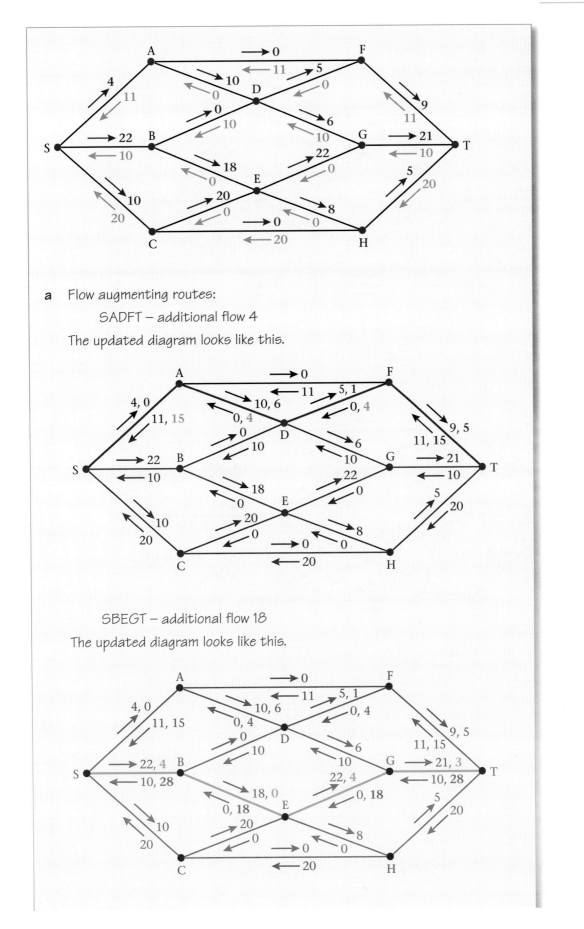

a Flow augmenting routes:

SADFT – additional flow 4

The updated diagram looks like this.

SBEGT – additional flow 18

The updated diagram looks like this.

SCEHT – additional flow 5

The updated diagram looks like this.

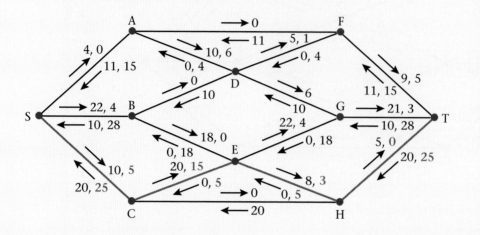

SCEGT – additional flow 3

The updated diagram looks like this.

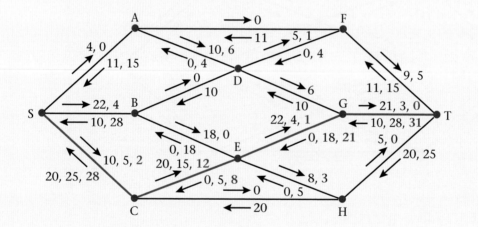

Look for a further route:

It cannot start SA since this arc has zero capacity in the direction SA.

It could start SB, but all the exits from B have zero capacity in the direction you would need to move, so SB is not possible.

It could start SC, then it must go along CE. You cannot go to H since HT has zero capacity and HC takes you back to C.

So you must start SCEG. From G you cannot proceed to T, but you could move to D.

So the route starts SCEGD and then proceeds to F and to T.

SCEGDFT – additional flow 1

The updated diagram looks like this.

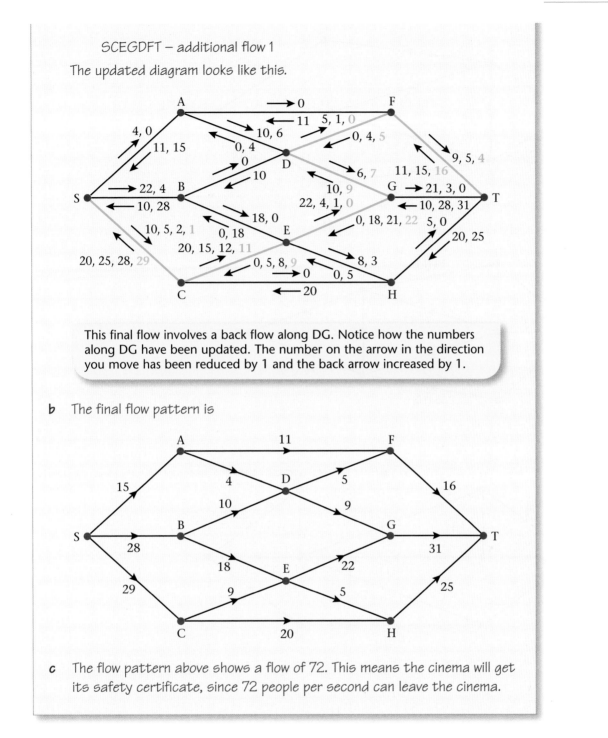

This final flow involves a back flow along DG. Notice how the numbers along DG have been updated. The number on the arrow in the direction you move has been reduced by 1 and the back arrow increased by 1.

b The final flow pattern is

c The flow pattern above shows a flow of 72. This means the cinema will get its safety certificate, since 72 people per second can leave the cinema.

Exercise 6D

There are photocopy masters on the CD-ROM for the questions in this exercise.

The diagrams show capacitated, directed networks. The capacity of each arc is shown on each arc. The numbers in circles represent an initial flow from S to T.

a Starting from the initial flow, use the labelling procedure to find a maximum flow through the network. You must list each flow-augmenting route you use together with its flow.

b Draw your final flow pattern and state the value of your maximum flow.

1

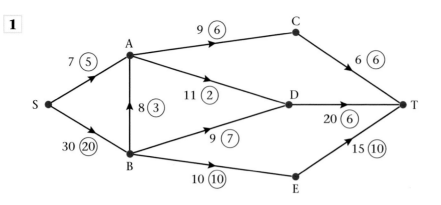

2

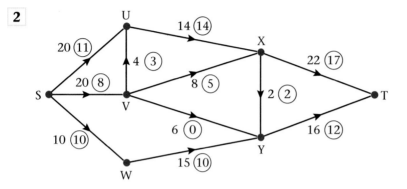

3

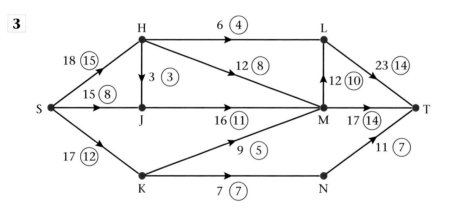

4

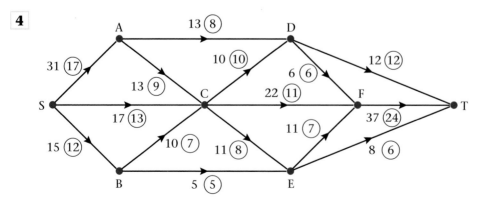

6.5 You can confirm that a flow is maximal using the maximum flow–minimum cut theorem.

The difficulty with the labelling procedure is that you may fail to spot a route. Therefore a method of checking whether the improved flow that you have found is maximal is useful.

The maximum flow–minimum cut theorem states:

■ **In a network the value of the maximum flow = the value of the minimum cut.**

> In passing from source to sink all of the flow must pass over the cut. A useful image is that of a border between two countries, A and B, with roads passing from A into B. The 'roads' are the arcs and the 'border' the cut. All traffic passing from A to B must pass over the border at some point and it could be used to restrict the number of vehicles entering country B. If only 500 vehicles per day are allowed to go from A to B, 501 cannot go through.
> Thus the minimum cut acts as restriction on the flow.

This means that you have a method of verifying that your flow is maximal.

■ **If you can find a cut with capacity equal to your flow, then the flow is maximal.**

To find the minimum cut you need to look for saturated arcs.

■ **The minimum cut passes through:**
> saturated arcs if directed from source to sink (into the cut),
> empty arcs if directed from sink to source (out of the cut).

> This does not mean that **every** saturated or empty arc must be in the cut, but it does mean that the cut can **only** pass through saturated arcs directed into the cut or empty arcs if directed out of the cut.

Example 12

(This is the network used in example 9.)

The diagram shows a capacitated, directed network and a flow of 38 passing through the same network.

Use the maximum flow–minimum cut theorem to prove that the flow is maximal.

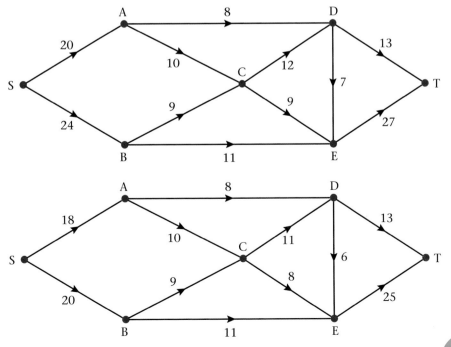

First identify the saturated arcs.

You do this by comparing the two diagrams. Saturated arcs will have a flow equal to their capacity. So you are looking for arcs which have the same number in both diagrams.

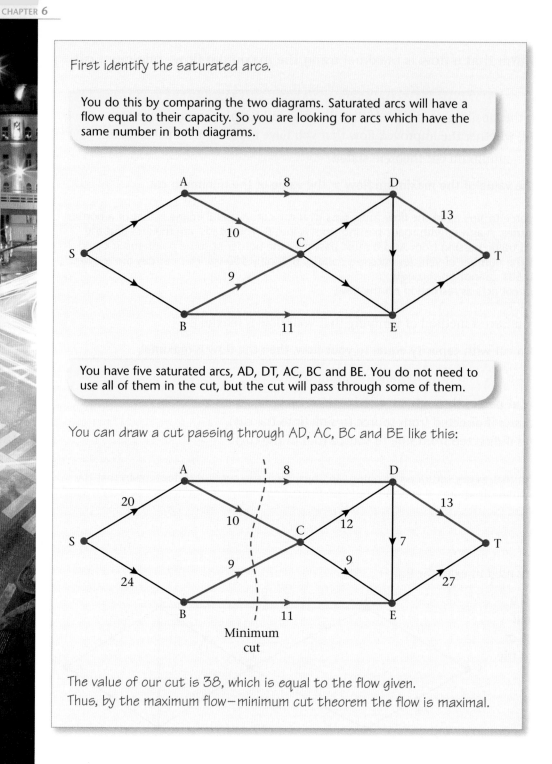

You have five saturated arcs, AD, DT, AC, BC and BE. You do not need to use all of them in the cut, but the cut will pass through some of them.

You can draw a cut passing through AD, AC, BC and BE like this:

The value of our cut is 38, which is equal to the flow given.
Thus, by the maximum flow—minimum cut theorem the flow is maximal.

Example 13

(This is the network used in example 10.)

The diagram shows a capacitated, directed network and a flow of 17 passing through the same network.

Use the maximum flow–minimum cut theorem to prove that the flow is maximal.

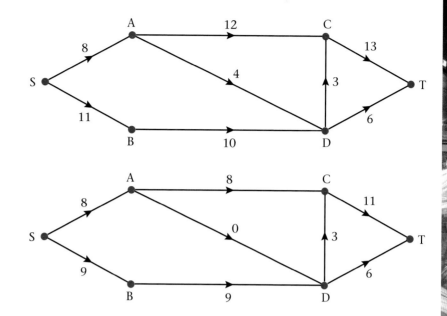

First identify the saturated arcs.

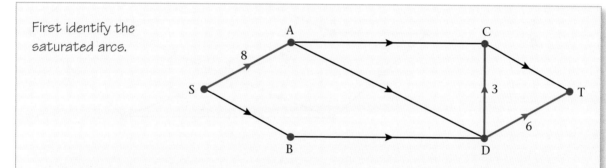

There are only three saturated arcs SA, DC and DT and they do not form a cut by themselves. However, notice that arc AD is empty, and that if it were included in the cut through SA, AD, DC and DT its direction would be 'out' of the cut, and so contribute zero to the capacity of the cut. Hence our cut passes through three saturated arcs (directed from the set of vertices containing the source to the set of vertices containing the sink) and one empty arc (directed from sink to source).

Draw a cut passing through SA, AD, DC and DT like this:

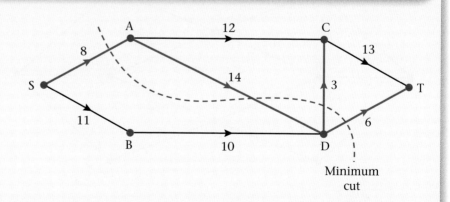

This cut has value 17, which is equal to the flow. Hence by the maximum flow–minimum cut theorem the flow is maximal.

Example 14

(This is the network used in example 11)

The diagram shows a capacitated, directed network and a flow of 72 passing thought the same network.

Use the maximum flow–minimum cut theorem to prove that the flow is maximal.

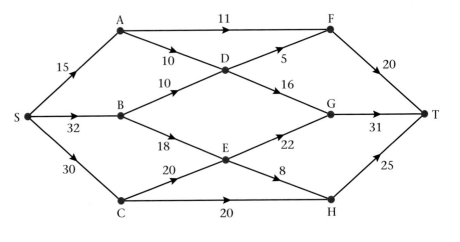

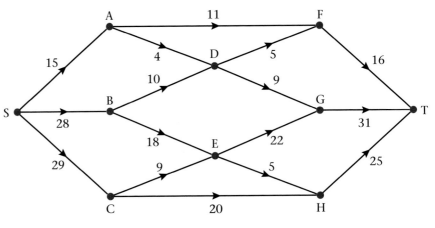

First identify the saturated arcs.

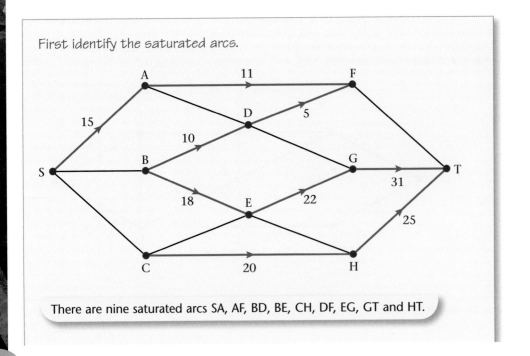

There are nine saturated arcs SA, AF, BD, BE, CH, DF, EG, GT and HT.

It is possible to find two minimum cuts like this:

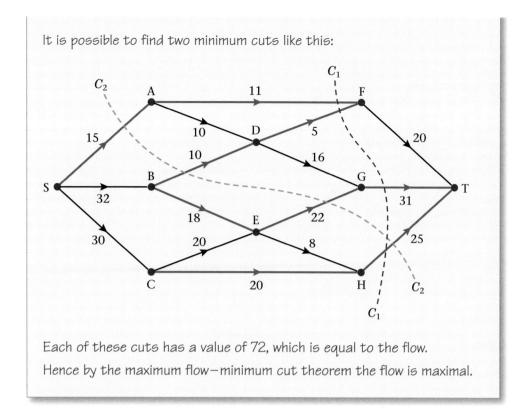

Each of these cuts has a value of 72, which is equal to the flow.

Hence by the maximum flow—minimum cut theorem the flow is maximal.

Exercise 6E

Use the maximum flow–minimum cut theorem to prove that the flows you found in answer to the questions in exercise 6D are maximal.

6.6 **You can adapt the algorithm to deal with networks with multiple sources and/or sinks.**

In networks with more than one source and/or sink, you create a **supersource** and/or a **supersink** together with arcs of appropriate capacity and then proceed as normal.

■ **The arcs leading from the supersource to each source, must have total capacity equal to the total capacity leaving each source.**

■ **The arcs leading from each of the sinks to the supersink must have capacity equal to that arriving at each sink.**

Example 15

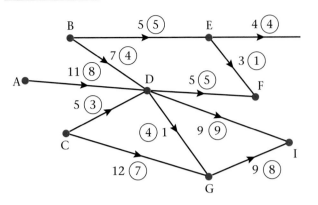

The diagram shows a capacitated, directed network The capacity of each arc is shown on each arc. The numbers in circles represent an initial flow.

a List the source vertices.

b List the sink vertices.

c Add supersource, S, a supersink, T, and appropriate arcs and flows to the diagram.

a The source vertices are A, B and C.

b The sink vertices are F, H and I.

c

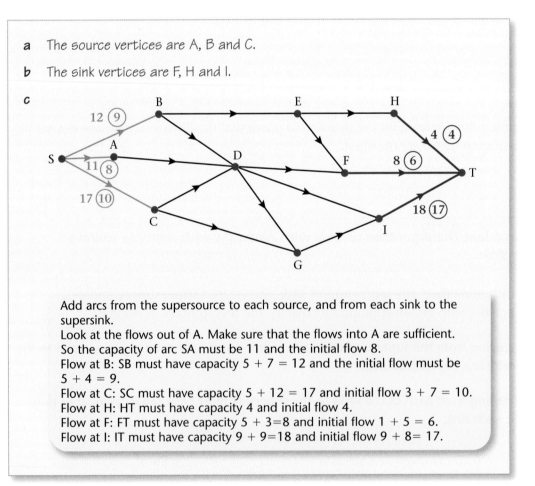

Add arcs from the supersource to each source, and from each sink to the supersink.
Look at the flows out of A. Make sure that the flows into A are sufficient.
So the capacity of arc SA must be 11 and the initial flow 8.
Flow at B: SB must have capacity 5 + 7 = 12 and the initial flow must be 5 + 4 = 9.
Flow at C: SC must have capacity 5 + 12 = 17 and initial flow 3 + 7 = 10.
Flow at H: HT must have capacity 4 and initial flow 4.
Flow at F: FT must have capacity 5 + 3 = 8 and initial flow 1 + 5 = 6.
Flow at I: IT must have capacity 9 + 9 = 18 and initial flow 9 + 8 = 17.

Example 16

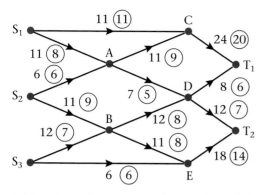

Water from three reservoirs, S_1, S_2 and S_3 is used to supply two towns T_1 and T_2, using a network of pipes.

The capacity of each pipe is given by the number on each arc. The numbers in circles represent an initial flow.

a Add a supersource, supersink and appropriate arcs to the diagram.

b Use the initial flow and the labelling procedure to find the maximum flow through the network. List each flow-augmenting route you use together with its flow.

c Draw your maximal flow pattern.

d State the value of your flow.

e Show that there are two minimum cuts.

a

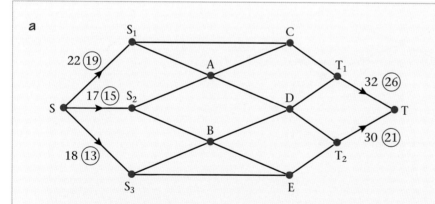

We add supersource, S, supersink, T, and arcs SS_1, SS_2, SS_3, T_1T and T_2T to the network..

Flow into S_1 must equal flow out of S_1, so the capacity is 22 and the initial flow 19.

Flow into S_2 must equal flow out of S_2, so the capacity is 17 and the initial flow 15.

Flow into S_3 must equal flow out of S_3, so the capacity is 18 and the initial flow 13.

Flow into T_1 must equal flow out of T_1, so the capacity is 32 and the initial flow 26.

Flow into T_2 must equal flow out of T_2, so the capacity is 30 and the initial flow 21.

b Apply the labelling procedure.

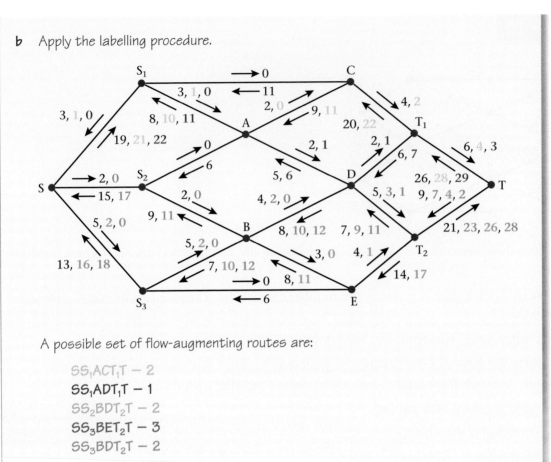

A possible set of flow-augmenting routes are:

$SS_1ACT_1T - 2$

$SS_1ADT_1T - 1$

$SS_2BDT_2T - 2$

$SS_3BET_2T - 3$

$SS_3BDT_2T - 2$

c

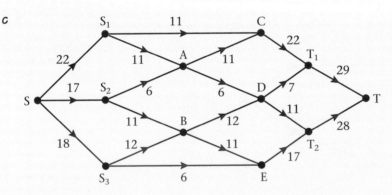

d The value of the flow is 57

e

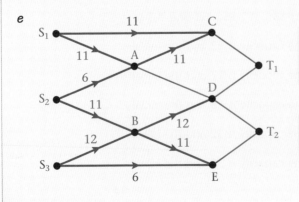

Note that you return to the original diagram to find the cuts. You cannot include in a cut any of the arcs you added to the diagram.

There are nine saturated arcs, where the current flow is equal to the capacity.

You can create two distinct cuts as shown.

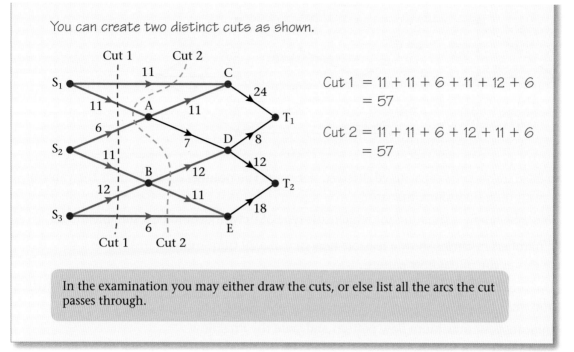

Cut 1 = 11 + 11 + 6 + 11 + 12 + 6
      = 57

Cut 2 = 11 + 11 + 6 + 12 + 11 + 6
      = 57

> In the examination you may either draw the cuts, or else list all the arcs the cut passes through.

It is acceptable in the examination to show your minimum cut on the same diagram as your final flow pattern, but you must use the **capacity** numbers in calculating the capacity of the cut.

Many candidates try to use the **flow** values.

The final flow pattern shows the final flow, so **all** 'cuts' through it will have the same value as the flow.

In order to find the minimum cut you **must** use the capacity numbers from the original diagram, but you can use the flow numbers to identify the saturated arcs.

Mixed exercise 6F

1

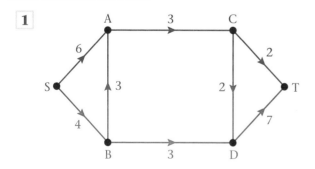

The diagram shows a capacitated, directed network. The number on each arc indicates the capacity of that arc.

a Use the labelling procedure to find the maximum flow through the network from S to T, listing each flow augmenting route you use, together with its flow.

b Verify that the flow found in part **a** is maximal.

2

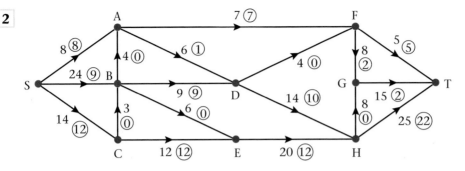

The diagram shows a capacitated directed network. The number on each arc is the value of the maximum flow along that arc.

a Describe briefly a situation for which this type of network could be a suitable model.

The numbers in circles show a feasible flow of value 29 from source S to sink T.
Take this as the initial flow pattern.

b Use the labelling procedure to find the maximum flow through the network from S to T. You must list each flow-augmenting route you use together with its flow.

c Indicate your maximum flow pattern and state the final flow.

d Verify that your answer is a maximum flow by using the maximum flow–minimum cut theorem, listing the arcs through which your cut passes.

e For the maximum flow, state a property of the arcs found in **d**. **E**

3

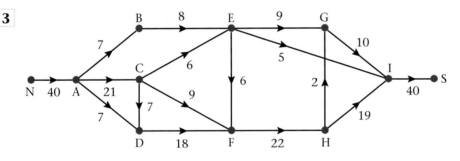

The diagram shows the road routes from a bus station, N, on the north side of a town to a bus station S, on its south side. The number on each arc shows the maximum flow rate, in vehicles per minute, on that route.

a State four junctions at which there could be traffic delays, giving a reason for your answer.

Given that AB, AD, CE, CF and EI are saturated,

b show a flow of 31 from N to S that satisfies this demand.

c Taking your answer to **b** as the initial flow pattern, use the labelling procedure to find the maximum flow. You should list each flow-augmenting route you use together with its flow.

d Indicate your maximum flow pattern.

e Verify your solution using the maximum flow–minimum cut theorem, listing the arcs through which your minimum cut passes.

f Show that, in this case, there is a second minimum cut and list the arcs through which it passes. **E**

4 A company wishes to transport its products from three factories F_1, F_2 and F_3 to a single retail outlet R. The capacities of the possible routes, in van loads per day, are shown.

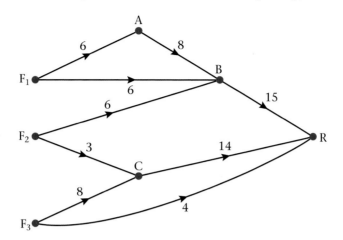

a On the worksheet add a supersource S to obtain a capacitated network with a single source and a single sink. State the minimum capacity of each arc you have added.

b i State the maximum flow along SF_1ABR and SF_3CR.

ii Show these maximum flows on the worksheet, using numbers in circles.

Taking your answer to part **b ii** as the initial flow pattern,

c i use the labelling procedure to find a maximum flow from S to R. List each flow-augmenting route you find together with its flow.

ii Prove that your final flow is maximal.

(E)

5 The network represents a road system through a town. The number on each arc represents the maximum number of vehicles that can pass along that road every minute, i.e. the capacity of the road.

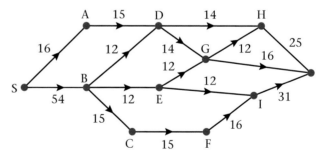

a State the maximum flow along
  i S B C F I T,   ii S A D H T.

b Show these maximum flows on a diagram.

c Taking your answer to part **b** as the initial flow pattern, use the labelling procedure to find a maximum flow from S to T. List each flow-augmenting route you find, together with its flow.

d Indicate a maximum flow.

e Prove that your flow is maximal.

The council has funding to improve one of the roads to increase the flow from S to T. It can choose to increase the flow along one of BE, DH or CF.

f Making your reasoning clear, explain which one of these three roads the council should improve, given that it wishes to maximise the flow through the town.

(E)

6 Figure 1 shows a capacitated, directed network. The number on each arc indicates the capacity of that arc.

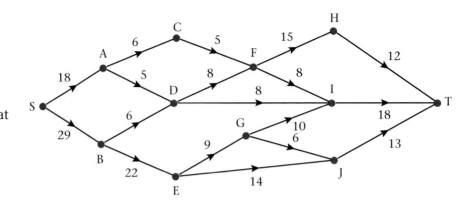

Figure 1

Figure 2 shows a feasible flow of value 29 through the same network.

a Find the values of the flows v, w, x, y and z.

Start with the values in Figure 1 and your answers to part **a** as your initial flow pattern.

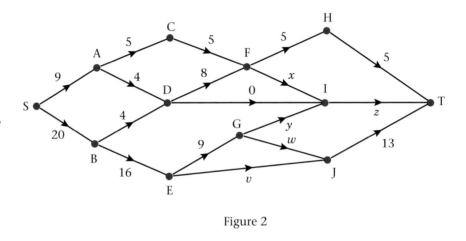

Figure 2

b Use the labelling procedure on Figure 1 to find the maximum flow through this network, listing each flow-augmenting route you use together with its flow.

c Show the maximum flow on Figure 2 and state its value.

d i Find the capacity of the cut which passes through the arcs HT, IT and JT.
  ii Find the minimum cut, listing the arcs through which it passes.
  iii Explain why this proves that the flow in part **c** is a maximum.

E

7

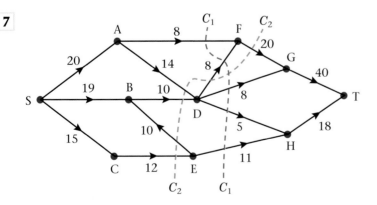

The diagram shows a capacitated, directed network. The number on each arc indicates the capacity of that arc.

a Calculate the values of cuts C_1 and C_2.

Given that one of these cuts is a minimum cut,

b state the maximum flow.

c Deduce the flow along GT, making your reasoning clear.

d By considering the flow into D, deduce that there are only two possible integer values for the flow along SA.

e For each of the two values found in part **d**, draw a complete maximum flow pattern.

f Given that the flow along each arc must be an integer, determine the number of other maximum flow patterns. Give a reason for your answer.

E

Summary of key points

1 In **capacitated directed networks**, (**capacitated directed graphs** or **capacitated digraphs**) each arc has a weight which represents the **capacity** of that arc.

2 The **feasibility condition** says that the flow along an arc must not exceed the capacity of that arc.

3 The **conservation condition** (on all but source and sink vertices) says that

   total flow into a vertex = total flow out of the vertex

4 If an arc contains a flow equal to its capacity it is **saturated**.

5 A **cut** is a set of arcs whose removal separates the network into two parts X and Y, where X contains at least the source and Y contains at least the sink.

6 The capacity (value) of a cut is the sum of the capacities of those arcs in the cut which are directed from X to Y.

7 When evaluating the capacity of a cut, only include the capacities of the arcs flowing **into** the cut.

8 The value of a cut is calculated using **capacities**.

9 In the labelling procedure, two arrows are drawn on each arc:
   - the forward arrow indicates any spare capacity
   - the backward arrow identifies by how much the flow in the arc could be reduced.

10 Each flow-augmenting route starts at S and finishes at T.

11 Forward and backward arrows can both be used as long as there is capacity in the direction in which you wish to move.

12 When sending an extra flow of value f along a flow-augmenting route, you

- subtract f from each 'forward' arrow on the route,
- add f to each 'backward' arrow on the route.

13 Sometimes it is necessary to make use of a backflow when finding a flow-augmenting route.

14 The maximum flow–minimum cut theorem states:

> in a network the value of the maximum flow = the value of the minimum cut

15 If you can find a cut, with capacity equal to the flow, then the flow is maximal.

16 The minimum cut passes through:

> saturated arcs if directed into the cut
> empty arcs if directed out of the cut.

17 In networks with more than one source and/or sink, you create a supersource and/or supersink together with arcs of appropriate capacity and then proceed as normal.

18 The arcs leading from the supersource to each source must have total capacity equal to the capacity leaving each source.

19 The arcs leading from each of the sinks to the supersink must have total capacity and total flow equal to that arriving at each sink.

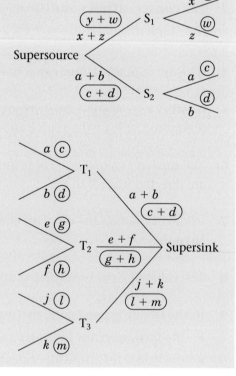

In this chapter you will:
- understand the terminology and principles of dynamic programming, including Bellman's principle of optimality
- be able to use dynamic programming to solve maximum and minimum problems, presented in network form
- be able to use dynamic programming to solve minimax and maximin problems, presented in network form
- be able to use dynamic programming to solve maximum, minimum, minimax or maximin problems, presented in table form.

Dynamic programming

Dynamic programming is a technique for solving multistage decision making problems. It finds optimal solutions without the use of calculus. Usually dynamic programming involves a sequence of decisions, the object being to optimise time, profit, cost, or resources by taking the correct decision at each stage. It was first developed by Richard Bellman and others in the 1950s as a management technique and has been applied to production planning, machine scheduling, stock control, allocation of resources, maintenance and replacement of equipment, investment planning and process design, amongst others.

7.1 You should understand the terminology and principles of dynamic programming, including Bellman's principle of optimality.

Some problems can be solved by drawing a directed network showing the possible decisions at each stage. You start at S and can make one of several decisions. Depending upon which one you take you can then take more decisions and so on until you eventually reach T.

The easiest problems are about minimising the cost or time. This is, of course, the same as finding the shortest path from S to T.

Bellman's Principle of Optimality

If the shortest/longest path from S to T is SABCT then

the shortest/longest path from S to C is SABC,

the shortest/longest path from S to B is SAB,

the shortest/longest path from S to A is SA,

and the shortest/longest path from A to C is ABC.

> You should already know this from your work on the shortest path algorithm. You are really just extending it to longest paths too.

- Any part of the shortest/longest path from S to T is itself a shortest/longest path.

> If there were a shorter/longer path between A and C, say, you would have chosen that route rather than ABC.

This rule works for

i the shortest/longest path **from S to any vertex** in the network and
ii between **two vertices both on the shortest/longest route**.

It **does not** apply to any two general vertices in the network.

- Bellman's principle of optimality states:

   any part of an optimal path is itself optimal

Some terminology

Stage   The route from the initial state (S say) to the final state (T say) is made up of a sequence of moves. Each move is a stage. The stage tells you how 'far' you are from the destination vertex.

State   This is the vertex that you are considering at each point.

Action   This is the directed arc from one state to the next. In selecting an arc you are considering what happens if you do that action.

Destination This is the vertex you arrive at having taken the action.

Value   This is the sum of the weights on the arcs used in a sequence of actions.

General principles of dynamic programming

When you use the shortest path algorithm from S to T, you **work forwards from S** through the network, finding the shortest route from S to every vertex.

■ In **dynamic programming** you **work backwards from T** in a series of stages.
  - The vertices immediately before T are examined. These are the stage 1 vertices. The best route from these to T is noted.
  - Then move back a stage, away from T and towards S, to the stage 2 vertices. Routes from these vertices must pass through one of the stage 1 vertices to get to T. Use this to help you find the optimal route from each stage 2 vertex to T.
  - Then move back a stage and repeat the process, until you reach S.
    The principle of optimality is used at each stage. The current optimal paths are developed using the paths found in the previous stage.

7.2 You should be able to use dynamic programming to solve maximum and minimum problems, presented in network form.

Example **1**

Use dynamic programming to find the minimum cost route from S to T in the network below.

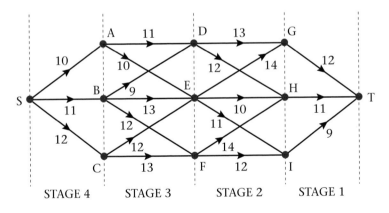

The solution is presented in the form of a table.

Stage	State	Action	Destination	Value
1	G	GT	T	12*
	H	HT	T	11*
	I	IT	T	9*
2	D	DG	G	13 + 12 = 25
		DH	H	12 + 11 = 23*
	E	EG	G	14 + 12 = 26
		EH	H	10 + 11 = 21
		EI	I	11 + 9 = 20*
	F	FH	H	14 + 11 = 25
		FI	I	12 + 9 = 21*

Begin at T so this is the first **stage**.
<u>Stage 1</u>
To get to T we must pass through G, H or I, so these are the **states**.
In each case there is no choice of route and so the only **actions** are GT, HT or IT.
The **values** of these actions are the weights of the arcs 12, 11 and 9 respectively.
So the states G, H, and I are given values of 12, 11 and 9 respectively.
This completes the first stage.

Use * to indicate the optimal action from each state.

Now move back a stage.

<u>Stage 2</u>
The stage 2 **states** are D, E and F. You need to find the shortest route from each vertex to T.
From D you must go to DG or DH. These are the only two **actions** from state D.
From E you must go to EG or EH or EI. These are the three **actions** from state E.
Finally from F, you must go to FH or FI, giving the two **actions** from state F.

You know the shortest route from each stage 1 vertex. You found them in the previous stage. To find the total length of your route so far, add the weights.

From D the actions and values are:
  action DG   value = 13 (DG) + 12 (GT) = 25
  action DH   value = 12 (DH) + 11 (HT) = 23
The smaller of these is 23 and this is indicated by *. This means that you can get from D to T in a minimum of 23.

From E the actions and values are:
  action EG   value = 14 (EG) + 12 (GT) = 26
  action EH   value = 10 (DH) + 11 (HT) = 21
  action EI   value = 11 (EI) + 9 (IT) = 20
The smallest of these is 20 and this is indicated by *. This means you can get from E to T in a minimum of 20.

From F the actions and values are:
  action FH   value = 14 (FH) + 11 (HT) = 25
  action FI   value = 12 (FI) + 9 (IT) = 21

The smaller of these is 21 and this is indicated by *.
This means you can get from F to T in a minimum of 21.

Now move back a stage.

3	A	AD	D	$11 + 23 = 34$
		AE	E	$10 + 20 = 30^*$
	B	BD	D	$9 + 23 = 32^*$
		BE	E	$13 + 20 = 33$
		BF	F	$12 + 21 = 33$
	C	CE	E	$12 + 20 = 32^*$
		CF	F	$13 + 21 = 34$
4	S	SA	A	$10 + 30 = 40^*$
		SB	B	$11 + 32 = 43$
		SC	C	$12 + 32 = 44$

Stage 3
The stage 3 **states** are A, B and C.
From A you must go to AD or AE. These are the only two **actions** from state A.
From B you must go to BD or BE or BF. These are the three **actions** from state B.
Finally from C, you must go to CE or CF, giving the two **actions** from state C.

From A the actions and values are:
    action AD   value = 11 (AD) + 23 (DT) = 34
    action AE   value = 10 (AE) + 20 (ET) = 30
The smaller of these is 30 and this is indicated by *. This means that you can get from A to T in a minimum of 30.

From B the actions and values are:
    action BD   value = 9 (BD) + 23 (DT) = 32
    action BE   value = 13 (BE) + 20 (ET) = 33
    action BF   value = 12 (BF) + 21 (FT) = 33
The smallest of these is 32 and this is indicated by *. This means you can get from B to T in a minimum of 32.

From C the actions and values are:
    action CE   value = 12 (CE) + 20 (ET) = 32
    action CF   value = 13 (CF) + 21 (FT) = 34
The smaller of these is 32 and this is indicated by *. This means that you can get from C to T in a minimum of 32.

Now move back a stage.

From S the actions and values are:
    action SA   value = 10 (SA) + 30 (AT) = 40
    action SB   value = 11 (SB) + 32 (BT) = 43
    action SC   value = 12 (SC) + 32 (CT) = 44
The smallest of these is 40 and this is indicated by *.

The shortest route is SAEIT of value 40.

To get the value of the route you look at the final stage and the optimal value is indicated by *.
To get the route, start at state S and look for the *. This is on SA, so the route begins SA.
Now look at state A for the *, this is on AE. So the route is SAE.
Now go to state E and look for the *. It is on EI, so the route is SAEI.
Finally look at state I for the *. It is on IT, so the route is SAEIT.

Example 2

Use dynamic programming to find the *maximum* cost route from S to T in the network below.

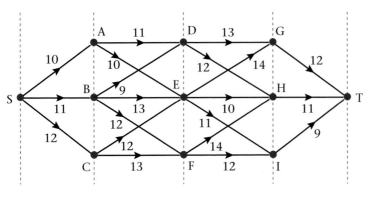

> Notice that the values that are * in each stage, are the ones you transfer to the next stage to calculate the value.

Stage	State	Action	Destination	Value
1	G	GT	T	12*
	H	HT	T	11*
	I	IT	T	9*
2	D	DG	G	13 + 12 = 25*
		DH	H	12 + 11 = 23
	E	EG	G	14 + 12 = 26*
		EH	H	10 + 11 = 21
		EI	I	11 + 9 = 20
	F	FH	H	14 + 11 = 25*
		FI	I	12 + 9 = 21
3	A	AD	D	11 + 25 = 36*
		AE	E	10 + 26 = 36*
	B	BD	D	9 + 25 = 34
		BE	E	13 + 26 = 39*
		BF	F	12 + 25 = 37
	C	CE	E	12 + 26 = 38*
		CF	F	13 + 25 = 38*
4	S	SA	A	10 + 36 = 46
		SB	B	11 + 39 = 50*
		SC	C	12 + 38 = 50*

> Note that each vertex is used as a state, and is used exactly once.

There are 3 maximum routes, all of length 50.

SBEGT, SCEGT and SCFHT

> To find the route, just 'follow the stars'.

Example **3**

Use dynamic programming to find

a the shortest,

b the longest,

route between S and T. State the route and its length in each case.

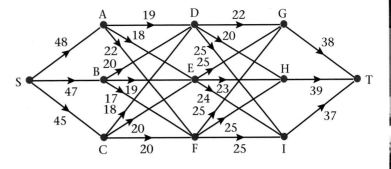

a   The shortest route

Stage	State	Action	Destination	Value
1	G	GT	T	38*
	H	HT	T	39*
	I	IT	T	37*
2	D	DG	G	22 + 38 = 60
		DH	H	20 + 39 = 59*
		DI	I	25 + 37 = 62
	E	EG	G	25 + 38 = 63
		EH	H	23 + 39 = 62
		EI	I	24 + 37 = 61*
	F	FG	G	25 + 38 = 63
		FH	H	25 + 39 = 64
		FI	I	25 + 37 = 62*
3	A	AD	D	19 + 59 = 78*
		AE	E	18 + 61 = 79
		AF	F	22 + 62 = 84
	B	BD	D	20 + 59 = 79
		BE	E	19 + 61 = 80
		BF	F	17 + 62 = 79*
	C	CD	D	18 + 59 = 77*
		CE	E	20 + 61= 81
		CF	F	20 + 62 = 82
4	S	SA	A	48 + 78 = 126
		SB	B	47 + 79 = 126
		SC	C	45 + 77 = 122*

The shortest route length is 122. The route is SCDHT.

The route length is indicated by the * in the final stage.
To find the route, 'follow the stars'. Start with SC, indicated by *, which sends you to C. The optimal route from C is CD (as indicated by *). This sends you to D. From state D you see that DH has been indicated by *, that sends you to H. Finally HT has been indicated by *. This gives you the route.

b The longest route

Stage	State	Action	Destination	Value
1	G	GT	T	38*
	H	HT	T	39*
	I	IT	T	37*
2	D	DG	G	22 + 38 = 60
		DH	H	20 + 39 = 59
		DI	I	25 + 37 = 62*
	E	EG	G	25 + 38 = 63*
		EH	H	23 + 39 = 62
		EI	I	24 + 37 = 61
	F	FG	G	25 + 38 = 63
		FH	H	25 + 39 = 64*
		FI	I	25 + 37 = 62
3	A	AD	D	19 + 62 = 81
		AE	E	18 + 63 = 81
		AF	F	22 + 64 = 86*
	B	BD	D	20 + 62 = 82*
		BE	E	19 + 63 = 82*
		BF	F	17 + 64 = 81
	C	CD	D	18 + 62 = 80
		CE	E	20 + 63 = 83
		CF	F	20 + 64 = 84*
4	S	SA	A	48 + 86 = 134*
		SB	B	47 + 82 = 129
		SC	C	45 + 84 = 129

Route length 134 and route SAFHT.

Example 4

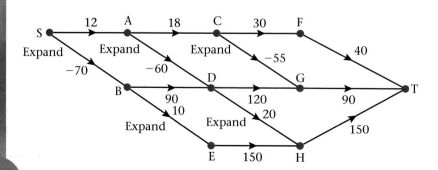

A firm is creating a three-year plan, and considering a possible expansion. If the firm expands there will be costs, due to purchasing new premises and equipment, but also increased revenue. To establish the effect of the decisions, the firm will then calculate the effect on realising all its assets at the end of the three-year period.

The firm can only expand a maximum of two times due to plant restrictions.

The results of the decisions are shown in the network above, with negative numbers indicating costs and positive numbers indicating profits, in £100 000.

Determine the strategy that maximises the profit over the three-year period.

This is a maximisation problem. You are seeking a maximum route from S to T.

Stage	State	Action	Destination	Value
1	F	FT	T	40*
(Realising assets)	G	GT	T	90*
	H	HT	T	150*
2	C	CF	F	30 + 40 = 70*
(Third year)		CG	G	−55 + 90 = 35
	D	DG	G	120 + 90 = 210*
		DH	H	20 + 150 = 170
	E	EH	H	150 + 150 = 300*
3	A	AC	C	18 + 70 = 88
(Second year)		AD	D	−60 + 210 = 150*
	B	BD	D	90 + 210 = 300
		BE	E	10 + 300 = 310*
4	S	SA	A	12 + 150 = 162
(First year)		SB	B	−70 + 310 = 240*

The maximum profit is £24 000 000.

There are units given in the question. In the final answer you express the value in terms of pounds rather than hundreds of thousands of pounds. It is a clearer statement of the profit for non-mathematicians to understand.

The route is SBEHT, which means:

Year	Year 1	Year 2	Year 3
Expand?	Expand	Expand	No expansion

So the firm should expand in both year 1 and year 2 (and cannot expand in year 3).

You are expected to be able to express your solutions, in practical terms, in such a way that a non-mathematician could follow the instructions.

Exercise 7A

Questions 1 to 4
Use dynamic programming to find

a a shortest

b a longest

route from S to T in each network below. State the route and its length in each case.

1

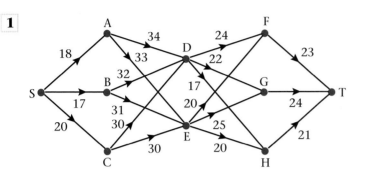

2

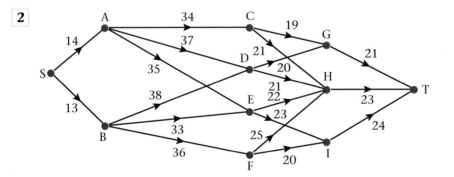

3

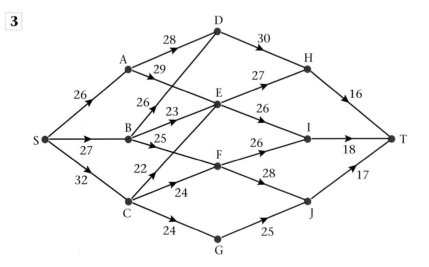

4

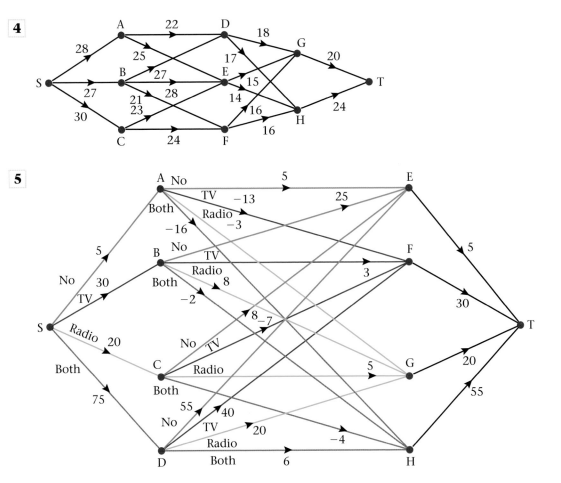

5

The diagram shows the effect on a company's profits, in £1000's, of taking various advertising decisions. The company wishes to create a two-year plan that will maximise its total profit.

Each year they must decide if they will not advertise (No), advertise through television only (TV), advertise through radio only (Radio), or advertise in both media (Both).

To determine the effectiveness of the strategy the company will estimate the value of its assets at the end of the two-year period.

Use dynamic programming to determine the advertising decisions that the directors should take.

7.3 You should be able to use dynamic programming to solve minimax and maximin problems, presented in network form.

You have already seen how to use dynamic programming to solve shortest and longest path (**minimum and maximum**) problems.

■ Sometimes you need to find a route where the longest arc is as small as possible (**minimax**), or a route where the smallest arc is as long as possible (**maximin**).

Imagine that the vertices in a network represent airports and the arcs represent distances between them. An aircraft is to carry cargo from S to T and you need to maximise the cargo that can be carried. To do this you need to minimise the fuel, and so the distance flown must also be minimised. You need a route where the longest leg is as short as possible. This is a **minimax** problem.

Imagine that the arcs represent the number of products processed per hour on a production line. The production rate will depend upon the slowest stage. You wish to find a route in which the smallest leg (slowest stage) is as big as possible. This would be a **maximin** problem.

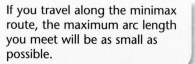

Use dynamic programming to find a minimax route from S to T.

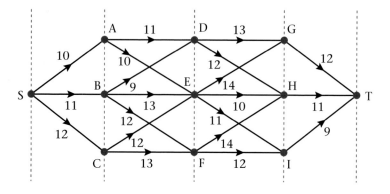

If you travel along the minimax route, the maximum arc length you meet will be as small as possible.

You are looking for the minimax route. So you need to record the maximum value of each action and select the minimum of these actions for each state.

So for **minimax**, **first** find the **maximums** and **then** the **minimum** of these.

Minimax

Stage	State	Action	Destination	Value
1	G	GT	T	12*
	H	HT	T	11*
	I	IT	T	9*
2	D	DG	G	Max (13, 12) = 13
		DH	H	Max (12, 11) = 12*
	E	EG	G	Max (14, 12) = 14
		EH	H	Max (10, 11) = 11*
		EI	I	Max (11, 9) = 11*
	F	FH	H	Max (14, 11) = 14
		FI	I	Max (12 , 9) = 12*
3	A	AD	D	Max (11, 12) =12
		AE	E	Max (10, 11) = 11*
	B	BD	D	Max (9, 12) = 12*
		BE	E	Max (13, 11) = 13
		BF	F	Max (12, 12) = 12*
	C	CE	E	Max (12, 11) = 12*
		CF	F	Max (13, 12) = 13
4	S	SA	A	Max (10, 11) = 11*
		SB	B	Max (11, 12) = 12
		SC	C	Max (12, 12) = 12

The minimax value is 11. There are two minimax routes: SAEHT and SAEIT.

Example 6

Use dynamic programming to
find a maximin route from S to T.

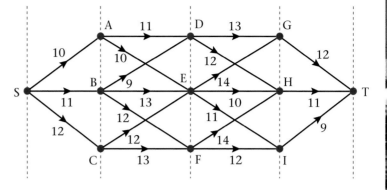

> This means that, if you travel along the maximin route, the minimum arc
> length you meet will be as large as possible.

You are looking for the maximin route. So you need to record the minimum value of each
action and select the maximum of these actions for each state.

So for **maximin**, **first** find the **minimums** and **then** the **maximum** of these.

Maximin

Stage	State	Action	Destination	Value
1	G	GT	T	12*
	H	HT	T	11*
	I	IT	T	9*
2	D	DG	G	Min (13, 12) = 12*
		DH	H	Min (12, 11) = 11
	E	EG	G	Min (14, 12) = 12*
		EH	H	Min (10, 11) = 10
		EI	I	Min (11, 9) = 9
	F	FH	H	Min (14, 11) = 11*
		FI	I	Min (12 , 9) = 9
3	A	AD	D	Min (11, 12) =11*
		AE	E	Min (10, 12) = 10
	B	BD	D	Min (9, 12) = 9
		BE	E	Min (13, 12) = 12*
		BF	F	Min (12, 11) = 11
	C	CE	E	Min (12, 12) = 12*
		CF	F	Min (13, 11) = 11
4	S	SA	A	Min (10, 11) = 10
		SB	B	Min (11, 12) = 11
		SC	C	Min (12, 12) = 12*

The maximin value is 12. The maximin route is SCEGT.

Exercise 7B

Questions 1 to 4

Use dynamic programming to find

a a minimax,

b a maximin,

route from S to T in each network below. State the route and its length in each case.

1

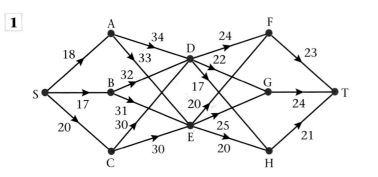

2

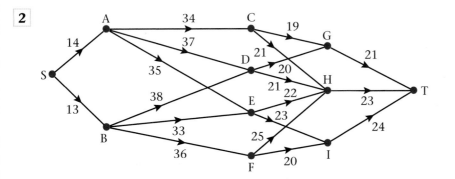

3

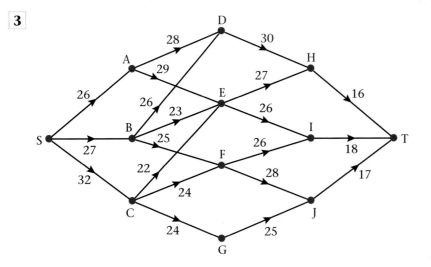

4

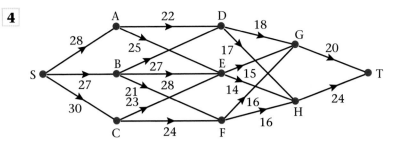

7.4 You should be able to use dynamic programming to solve maximum, minimum, minimax or maximin problems, presented in table form.

Example 7

A fruit grower can use his surplus crop to make into jam, tinned pie filling or dried fruit.

He can process up to five units of fruit and the expected profits, in hundreds of pounds, from making the various products are shown in the table. He will use all five units to make these products and wishes to maximise his income.

Number of units of fruit	1	2	3	4	5
Jam	12	27	33	36	38
Tinned pie filling	16	18	20	22	24
Dried fruit	8	16	24	32	40

Dynamic programming will be used to solve this problem.

a Define the terms Stage, State, Action, Destination and Value in this context.

b Determine the number of units he should allocate to each product in order to maximise his income.

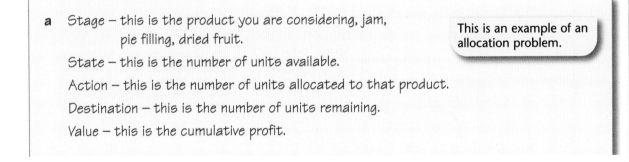

a Stage – this is the product you are considering, jam, pie filling, dried fruit.

State – this is the number of units available.

Action – this is the number of units allocated to that product.

Destination – this is the number of units remaining.

Value – this is the cumulative profit.

> This is an example of an allocation problem.

b

Stage (product)	State (units available)	Action (units allocated)	Destination (units remaining)	Value (cumulative profit)
1	0	0	0	0*
(Dried	1	1	0	8*
Fruit)	2	2	0	16*
	3	3	0	24*
	4	4	0	32*
	5	5	0	40*
2	0	0	0	0 + 0 = 0*
Pie	1	1	0	16 + 0 = 16*
Filling		0	1	0 + 8 = 8
	2	2	0	18 + 0 = 18
		1	1	16 + 8 = 24*
		0	2	0 + 16 = 16
	3	3	0	20 + 0 = 20
		2	1	18 + 8 = 26
		1	2	16 + 16 = 32*
		0	3	0 + 24 = 24
	4	4	0	22 + 0 = 22
		3	1	20 + 8 = 28
		2	2	18 + 16 = 34
		1	3	16 + 24 = 40*
		0	4	0 + 32 = 32
	5	5	0	24 + 0 = 24
		4	1	22 + 8 = 30
		3	2	20 + 16 = 36
		2	3	18 + 24 = 42
		1	4	16 + 32 = 48*
		0	5	0 + 40 = 40
3	5	5	0	38 + 0 = 38
Jam		4	1	36 + 16 = 52
		3	2	33 + 24 = 57
		2	3	27 + 32 = 59*
		1	4	12 + 40 = 52
		0	5	0 + 48 = 48

You are working backwards, so you are considering the last product first. Since this is the last product, all the remaining surplus fruit must be used, so all the destination values must be zero. If there are n units left at this stage all n must be allocated to dried fruit.

We now move back a stage to the next product.

If there is one unit available then you can either allocate one unit to pie filling leaving no units for dried fruit, or no units for pie filling leaving one left for dried fruit.

If there are two units available you can allocate two, one or none to pie filling, leaving none, one or two left for dried fruit.
If you allocate 2 to pie filling the profit is 18 and you add 0 because you have no units left to allocate to dried fruit.
If you allocate 1 to pie filling the profit is 16 and you add 8 because you will use the remaining unit for dried fruit.
If you allocate no units to pie filling the profit is 0 but you add 16 because you will use the two units for dried fruit.

If there are three units available you can allocate three, two, one or none to pie filling, leaving none, one, two or three left for dried fruit.
If you allocate 3 to pie filling the profit is 20 and you add 0 because you have no units left to allocate to dried fruit.
If you allocate 2 to pie filling the profit is 18 and you add 8 because you will use the remaining units for dried fruit.
If you allocate 1 to pie filling the profit is 16 and you add 16 because you will use the remaining units for dried fruit.
If you allocate none to pie filling the profit is 0 and you add 24 because you will use the remaining units for dried fruit.

To obtain the maximum profit 2 units should be allocated to Jam, 1 to pie filling and 2 to dried fruit. This will give a profit of £5900.

Example 8

A clockmaker makes grandfather clocks during the winter. The order book for clocks over the next five months is shown in the table below.

Month	October	November	December	January	February
Number of clocks ordered	2	3	5	3	1

The clocks are delivered to customers at the end of each month.

He can make up to four in any month, but if he needs to make more than two in any one month he will need to hire additional help at £500 per month.

The overhead costs are £300 in any month in which work is done.

He can put up to two clocks into storage at a cost of £100 per clock per month.

There are no clocks in storage at the beginning of October and there should be no clocks in storage after the February delivery.

Use dynamic programming to determine the production schedule that minimises the costs.

> This is an example of a production planning problem.

Stage (month)	Demand (Number of clocks to be delivered at the end of the month)	State (Number of clocks in storage at start of month)	Action (Number of clocks made in the month)	Destination (Number of clocks to put in storage at end of month)	Value (production costs)
1	1	0	1	0	$300 = 300^*$
Feb		1	0	0	$100 = 100^*$
2	3	0	3	0	$500 + 300 + 300 = 1100$
Jan			4	1	$500 + 300 + 100 = 900^*$
		1	2	0	$300 + 100 + 300 = 700^*$
			3	1	$500 + 300 + 100 + 100 = 1000$
		2	1	0	$300 + 200 + 300 = 800$
			2	1	$300 + 200 + 100 = 600^*$
3	5	1	4	0	$500 + 300 + 100 + 900 = 1800^*$
Dec		2	3	0	$500 + 300 + 200 + 900 = 1900$
			4	1	$500 + 300 + 200 + 700 = 1700^*$

4	3	0	4	1	**500** + **300** + **1800** = **2600***
Nov		1	3	1	**500** + **300** + **100** + **1800** = **2700**
			4	2	**500** + **300** + **100** + **1700** = **2600***
		2	2	1	**300** + **200** + **1800** = **2300***
			3	2	**500** + **300** + **200** + **1700** = **2700**
5	2	0	2	0	**300** + **2600** = **2900***
Oct			3	1	**500** + **300** + **2600** = **3400**
			4	2	**500** + **300** + **2300** = **3100**

> An extra column has been added, 'Demand', to help with the planning.

> Some notes of explanation have been added to the column headings, so that you are clear about the meaning of each term in this context. You may be asked to state these in the examination, but even if you are not asked to, it is wise to clarify their meanings.

So the clockmaker should make the clocks as follows

Month	October	November	December	January	February
Number of clocks made	2	4	4	4	0

The costs will be £2 900.

Remarks for table above:

Start by filling in the first five columns:

In February: You require 1 clock and none in storage at the end of this final month. So at most there should be only one clock in storage at the start of the month.

In February: Since you need just one clock and none for storage, you can work out the possible actions and destinations.

In January: You require 3 clocks and can put at most one into storage (since you can only have one in storage in February). You could have up to two clocks in storage

In January: If you have no clocks in store you must make 3, to meet the demand, and could make 4.

If you have one clock in store you must make 2, to meet the demand, and could make 3 putting one in store, but no more, because you can only have one in store for February.

If you have 2 clocks in store you must make one, to meet demand, and you could make 2, putting one in store, but no more because you can only have one in store for February.

In December: You require 5 clocks, but can only make 4 at maximum in the month, so you must have at least one in store and could have 2.

In December: If you have 1 in store you must make 4 to meet the demand.

If you have 2 in store you must make 3 to meet the demand, and could make 4, putting one in store.

In November: You require 3 clocks, but must have at least one extra to put in store for December. This means you need to have at least 4 clocks by the end of the month. You could therefore have either none in store, or 1 or 2.

In November: If you have none in store you must make 4 (3 to meet demand and 1 to put in store).

If there is 1 in store you must make at least 3, and could make 4, putting either one or two in store.

If you have 2 in store you must make 2, and could make 3, putting 1 or 2 in store.

In October: You need 2 and have none in store.

In October: You must make at least 2 to meet the demand and could make up to 2 more, putting either 1 or 2 in store.

To work out the costs you use the information given:

If you need to make more than two in any one month you will need to hire additional help at £500 per month.

The overhead costs are £300 in any month in which work is done.

You can put up to two clocks into storage at a cost of £100 per clock per month.

Remember that the optimal values found for each state must be carried into the stage below.

Example 9

A builder has purchased a site and will build three houses A, B, C on it at the rate of one a year. As each house is built it causes access problems for future work, and construction costs will rise each year. The builder's estimates of costs are shown in the table below, in thousands of pounds.

Already built	A	B	C
Nothing	70	80	75
A	–	90	80
B	90	–	95
C	95	100	–
A and B	–	–	105
A and C	–	110	–
B and C	115	–	–

For tax reasons the builder needs to choose the order in which he builds each house so that the least annual cost is as large as possible.

Dynamic programming will be used to determine the order in which the houses should be built.

a Explain the meaning of Stage, State, Action in this context.

b Find the order in which the houses should be built and the least annual cost incurred.

a   Stage — the number of houses left to build
    State — the houses already built
    Action — the house to be built.

b

Stage	State	Action	Destination	Value
1	AB	C	ABC	105*
	AC	B	ABC	110*
	BC	A	ABC	115*
2	A	B	AB	Min (90, 105) = 90*
		C	AC	Min (80, 110) = 80
	B	A	AB	Min (90, 105) = 90
		C	BC	Min (95, 115) = 95*
	C	A	AC	Min (95, 110) = 95
		B	BC	Min (100, 115) = 100*
3	None	A	A	Min (70, 90) = 70
		B	B	Min (80, 95) = 80*
		C	C	Min (75, 100) = 75

You need to maximise the least annual cost so this is a maximin problem.

The optimal order is B, C, A and the maximin value is 80 000.

Example 10

Toby has a fairground ride and will take it to three county fairs over the next three weeks.

He will leave home at the start of week 1 and travel to the first fair, then go to the second fair for the second week and then go to the third fair for the final week before returning home.

There are seven fairs during the next three weeks so there is a choice of fair each week. Toby must decide which one of the fairs he should go to.

Table 1 gives the week in which each fair is held. Table 2 gives the expected profits, in hundreds of pounds, in going to each fair. Table 3 gives the travel costs in hundreds of pounds.

Table 1

Week	1	2	3
Fair	A, B	C, D, E	F, G

Table 2

Fair	A	B	C	D	E	F	G
Profit (£100)	6	5	8	10	11	14	12

Table 3

Travel costs (£100)	A	B	C	D	E	F	G
Home	3	2				8	7
A			3	2	1		
B			1	3	3		
C						5	5
D						4	2
E						4	1

Toby wishes to maximise his income.

Use dynamic programming to determine the fairs he should go to and his total income.

First define the stage, state and action.

Stage – the number of weeks remaining.

State – Toby's location during week.

Action – the journey that Toby will make at the end of the week.

You need also to decide the value calculation. In this example the value will be the profit gained at the fair just attended minus the cost of the journey. This means that in stage 4 you will just have travel costs.

This is a maximisation problem.

Stage	State	Action	Destination	Value
1	F	F-Home	Home	$14 - 8 = 6^*$
	G	G-Home	Home	$12 - 7 = 5^*$
2	C	CF	F	$8 - 5 + 6 = 9^*$
		CG	G	$8 - 5 + 5 = 8$
	D	DF	F	$10 - 4 + 6 = 12$
		DG	G	$10 - 2 + 5 = 13^*$
	E	EF	F	$11 - 4 + 6 = 13$
		EG	G	$11 - 1 + 5 = 15^*$
3	A	AC	C	$6 - 3 + 9 = 12$
		AD	D	$6 - 2 + 13 = 17^*$
		AE	E	$6 - 5 + 15 = 16$
	B	BC	C	$5 - 1 + 9 = 13$
		BD	D	$5 - 3 + 13 = 15$
		BE	E	$5 - 3 + 15 = 17^*$
4	Home	Home-A	A	$-3 + 17 = 14$
		Home-B	B	$-2 + 17 = 15^*$

The value here is made up of the profit from attending E, minus the travel cost from E to F, plus the value found from state F.

Toby should attend fairs B, E and G. He will make an income of £15 000.

1 A company is created to sell holidays on an island. There are three new resorts A, B and C being created on the island and the company decides to introduce one new resort to its catalogue each year over the next three years. The costs of introducing each resort will be influenced by the number of resorts listed in the catalogue. The more resorts the company has listed, the smaller the cost of adding another resort. The estimates of annual costs are shown in the table below, in hundreds of pounds.

Resorts listed	A	B	C
None	60	60	55
A	–	50	60
B	40	–	55
C	35	50	–
A and B	–	–	50
A and C	–	45	–
B and C	30	–	–

For funding reasons the company needs to choose the order in which the resorts are introduced so that the greatest annual cost is as small as possible.

Dynamic programming will be used to determine the order in which the resorts are introduced.

a Explain the meaning of Stage, State and Action in this context.

b Find the order in which the resorts should be added and the greatest annual cost.

2 A house renovation project is to be completed in six weeks (30 working days). The work is in four phases: clearance, repairing, modernisation and decorating, which must be undertaken in that order. The cost, in £1000, of each stage depends on the time taken to do it. These are shown in the table.

Time for stage (days)	Clearance	Repairing	Modernisation	Decorating
5	15	24	22	14
10	13	20	19	12
15	8	15	15	9
20	5	10	11	4
25	2	6	7	2

Dynamic programming will be used to solve this problem.

a Define the terms Stage, State, Action, Destination and Value in this context.

b Determine the number of days that should be allocated to each stage in order to minimise costs.

3 A company makes aircraft. The order book over the next four months is shown in the table below.

Month	March	April	May	June
Number of aircraft ordered	1	2	3	2

The aircraft are delivered to customers at the end of each month.

Up to three aircraft can be made in any month, but if more than two are made in any one month additional equipment will need to be hired at £20 000 per month.

If any work is done in a month the overhead costs are £50 000.

Up to two aircraft can be held in secure hangers at a cost of £10 000 per aircraft per month.

There are no aircraft in store at the beginning of March and there should be no aircraft in store after the June delivery.

Use dynamic programming to determine the production schedule that minimises the costs.

4 A salesman will visit four shops in the next four days to demonstrate a new product. He will start at home and travel to the first shop and spend the first day there, then travel directly to the second shop for day 2, onto the third shop for day 3, then to the fourth shop for day 4 and then travel home.

Table 1 shows the shops he could visit on each day.
Table 2 shows the anticipated profit, in £100, from sales at each shop.
Table 3 shows the travelling expenses, in £100, that will be incurred.

The company employing the salesman wishes to maximise the income, after subtracting the travel costs, generated by the salesman's visits. Find his optimum route.

Table 1

Monday	Tuesday	Wednesday	Thursday
A, B, C	D, E	F, G	H, I, J

Table 2

Shop	A	B	C	D	E	F	G	H	I	J
Profit	8	9	8	12	14	10	11	14	13	11

Table 3

	A	B	C	D	E	F	G	H	I	J
Home	2	2	3					6	4	3
A				3	4					
B				4	6					
C				4	4					
D						5	5			
E						4	7			
F								5	4	4
G								5	5	4

Mixed exercise 7D

1

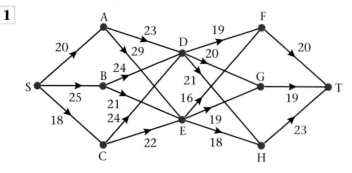

Use dynamic programming to find the minimax route from S to T in the network above.

2 Using the same diagram as in question 1, use dynamic programming to find the maximin route.

3 A dairy manufacturer can make butter, cheese and yoghurt. Up to five units of milk can be processed and the profits from the various allocations are shown in the table.

Number of units	1	2	3	4	5
Butter	14	25	34	41	47
Cheese	12	30	40	45	49
Yoghurt	10	20	30	40	50

The manufacturer wishes to maximise his profit.

a Use dynamic programming to find an optimal solution and state the profit.

b Show that there is a second optimal solution.

4 Jenny wishes to travel from S to T. There are several routes available. She wishes to choose the route on which the maximum altitude, above sea level, is as small as possible. This is called the minimax route.

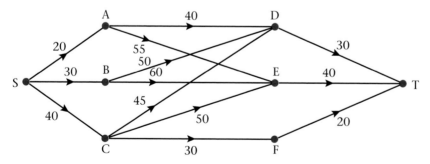

The diagram gives the possible routes and the weights on the edges give the maximum altitude on the road (in units of 100 feet).

Use dynamic programming, carefully defining the stages and states, to determine the route or routes Jenny should take. You should show your calculations in tabular form, using a table with columns labelled as shown below.

Stage	Initial state	Action	Final state	Value

E

5 At the beginning of each month an advertising manager must choose one of three adverts:

>the previous advert;
>the current advert;
>a new advert.

She therefore has 3 options:

>A: use the previous advert;
>B: use the current advert;
>C: run a new advert.

The possible choices are shown in the network below together with (stage, state) variables at the vertices and the expected profits, in thousands of pounds, on the arcs.

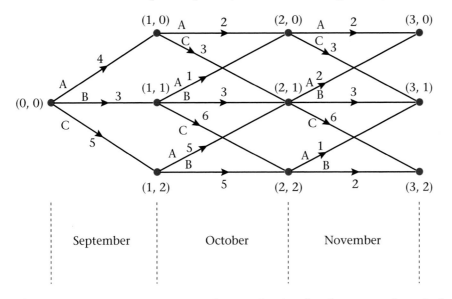

The manager wants to maximise her profits for the three-month period.

a Complete the table on the worksheet.

b Hence obtain the sequence of decisions she should make to obtain the maximum profit. State the maximum profit. **E**

Summary of key points

1   A **stage** is a move along the path from the initial state to the final state.

2   A **state** is the vertex being considered.

3   An **action** is a directed arc from one state to the next.
    In selecting an arc you consider what happens if you do that action.

4   A **destination** is the vertex you arrive at having taken an action.

5   A **value** is the sum of the weights on the arcs used in a sequence of actions.

6   Bellman's principle of optimality states:

    any part of an optimal path is itself optimal.

7   In **dynamic programming** you work backwards from the final destination, T, in a series
    of stages, to the initial vertex, S.

8   At each stage you choose the optimal route. The current optimal paths are developed using
    the paths found in the stage before.

9   You can use dynamic programming to solve maximum and minimum problems presented
    in network form.

10 A **minimax** problem is one in which the longest leg is as short as possible.

11 A **maximin** problem is one in which the shortest leg is as long as possible.

12 You can use dynamic programming to solve minimax and maximin problems presented in
    network and table form.

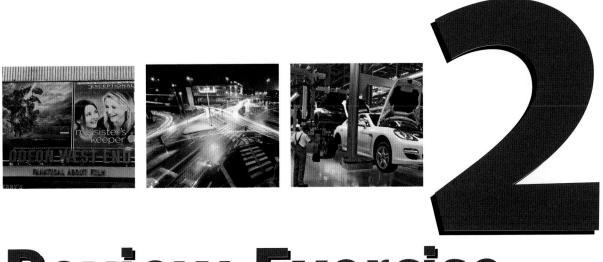

Review Exercise

2

Photocopy masters are available on the CD-ROM for questions marked *.

1 In a game theory explain what is meant by

a zero-sum game,

b saddle point. **E**

2

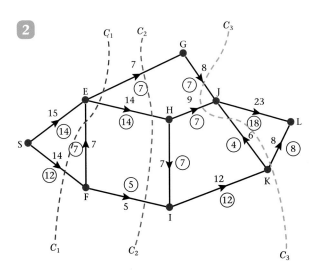

The diagram shows a network of roads represented by arcs. The capacity of the road represented by that arc is shown on each arc. The numbers in circles represent a possible flow of 26 from S to L.

Three cuts C_1, C_2 and C_3 are shown.

a Find the capacity of each of the three cuts.

b Verify that the flow of 26 is maximal.

The government aims to maximise the possible flow from S to L by using one of two options.

> Option 1: Build a new road from E to J with capacity 5.

or

> Option 2: Build a new road from F to H with capacity 3.

c By considering **both** options, explain which one meets the government's aim. **E**

3 A two person zero-sum game is represented by the following pay-off matrix for player A.

	B plays I	B plays II	B plays III
A plays I	−3	2	5
A plays II	4	−1	−4

a Write down the pay-off matrix for player B.

b Formulate the game as a linear programming problem for player B, writing the constraints as equalities and stating your variables clearly. **(E)**

4 An engineering firm makes motors. They can make up to five in any one month, but if they make more than four they have to hire additional premises at a cost of £500 per month. They can store up to two motors for £100 per motor per month. The overhead costs are £200 in any month in which work is done.

Motors are delivered to buyers at the end of each month. There are no motors in stock at the beginning of May and there should be none in stock after the September delivery.

The order book for motors is:

Month	May	June	July	Aug.	Sept.
Number of motors	3	3	7	5	4

Use dynamic programming to determine the production schedule that minimises the costs, showing your working in a table. **(E)**

5

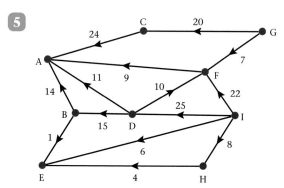

The diagram shows a capacitated directed network. The number on each arc is its capacity.

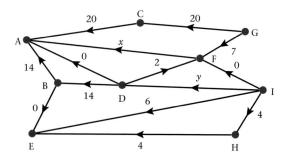

This shows a feasible initial flow through the same network.

a Write down the values of the flow x and the flow y.

b Obtain the value of the initial flow through the network, and explain how you know it is not maximal.

c Use this initial flow and the labelling procedure to find a maximum flow through the network. You must list each flow-augmenting route you use, together with its flow.

d Show your maximal flow pattern.

e Prove that your flow is maximal. **(E)**

6 A two-person zero-sum game is represented by the following pay-off matrix for player A.

		B			
		I	II	III	IV
	I	-4	-5	-2	4
A	II	-1	1	-1	2
	III	0	5	-2	-4
	IV	-1	3	-1	1

a Determine the play safe strategy for each player.

b Verify that there is a stable solution and determine the saddle points.

c State the value of the game to B. **(E)**

7

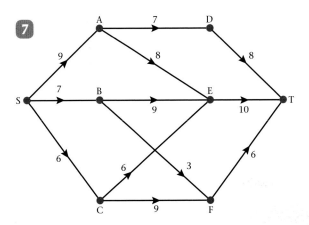

The network shows possible routes that an aircraft can take from S to T. The numbers on the directed arcs give the amount of fuel used on that part of the route, in appropriate units. The airline wishes to choose the route for which the maximum amount of fuel used on any part of the route is as small as possible. This is the minimax route.

a Complete a table to show the information.

b Hence obtain the minimax route from S to T and state the maximum amount of fuel used on any part of this route. **E**

8 *

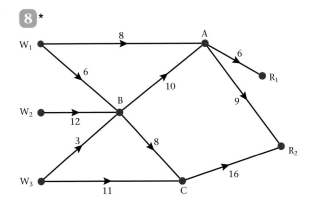

A company has three warehouses W_1, W_2 and W_3. It needs to transport the goods stored there to two retail outlets R_1 and R_2. The capacities of the possible routes, in van loads per day, are shown. Warehouses W_1, W_2 and W_3 have 14, 12 and 14 van loads respectively available per day and retail outlets R_1 and R_2 can accept 6 and 25 van loads respectively per day.

a On a copy of the diagram add a supersource W, a supersink R and the appropriate directed arcs to obtain a single-source, single-sink capacitated network. State the minimum capacity of each arc you have added.

b State the maximum flow along
   i W W_1 A R_1 R,
   ii W W_3 C R_2 R.

c Taking your answers to part **b** as the initial flow pattern, use the labelling procedure to obtain a maximum flow through the network from W to R. Show your working. List each flow-augmenting route you use, together with its flow.

d From your final flow pattern, determine the number of van loads passing through B each day.

The company has the opportunity to increase the number of van loads from one of the warehouses W_1, W_2, W_3 to A, B or C.

e Determine how the company should use this opportunity so that it achieves a maximum flow. **E**

9 Emma and Freddie play a zero-sum game. This game is represented by the following pay-off matrix for Emma.

$$\begin{pmatrix} -4 & -1 & 3 \\ 2 & 1 & -2 \end{pmatrix}$$

a Show that there is no stable solution.

b Find the best strategy for Emma and the value of the game to her.

c Write down the value of the game to Freddie and his pay-off matrix. **E**

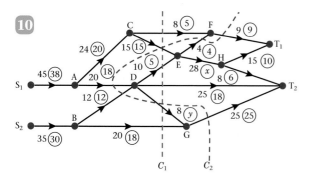

10

The diagram shows a capacitated, directed network. The unbracketed number on each arc indicates the capacity of that arc, and the numbers in circles show a feasible flow of value 68 through the network.

a Add a supersource and a supersink, and arcs of appropriate capacity, to a copy of the diagram.

b Find the values of x and y, explaining your method briefly.

c Find the value of cuts C_1 and C_2.

Starting with the given feasible flow of 68,

d use the labelling procedure to find a maximal flow through this network. List each flow-augmenting route you use, together with its flow.

e Show your maximal flow and state its value.

f Prove that your flow is maximal. **E**

11 *a** Explain what is meant by a maximin route in dynamic programming, and give an example of a situation that would require a maximin solution.

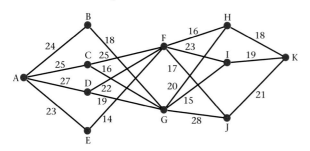

A maximin route is to be found through the network shown.

b Complete the table on the worksheet, and hence find a maximin route.

c List **all** other maximin routes through the network. **E**

12 *

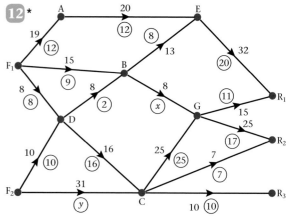

The diagram shows a capacitated, directed network of pipes flowing from two oil fields F_1 and F_2 to three refineries R_1, R_2 and R_3. The number on each arc represents the capacity of the pipe and the numbers in the circles represent a possible flow of 65.

a Find the value of x and the value of y.

b On the worksheet, add a supersource and a supersink, and arcs showing their minimum capacities.

c Taking the given flow of 65 as the initial flow pattern, use the labelling procedure to find the maximum flow. State clearly your flow-augmenting routes.

d Show the maximum flow and write down its value.

e Verify that this is the maximum flow by finding a cut equal to the flow. **E**

13 A two person zero-sum game is represented by the following pay-off matrix for player A.

	B plays I	B plays II	B plays III
A plays I	2	−1	3
A plays II	1	3	0
A plays III	0	1	−3

a Identify the play safe strategies for each player.

b Verify that there is no stable solution to this game.

c Explain why the pay-off matrix above may be reduced to

	B plays I	B plays II	B plays III
A plays I	2	−1	3
A plays II	1	3	0

d Find the best strategy for player A, and the value of the game. ⓔ

14 *

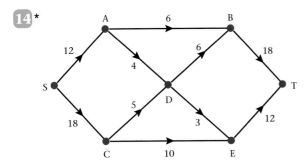

The diagram shows a capacitated network. The numbers on each arc indicate the capacity of that arc in appropriate units.

a Explain why it is not possible to achieve a flow of 30 through the network from S to T.

b State the maximum flow along
i SABT, **ii** SCET.

c Show these flows on the worksheet.

d Taking your answer to part **c** as the initial flow pattern, use the labelling procedure to find a maximum flow from S to T. Show your working. List each flow-augmenting path you use together with its flow.

e Indicate a maximum flow.

f Prove that your flow is maximal. ⓔ

15 Andrew (A) and Barbara (B) play a zero-sum game. This game is represented by the following pay-off matrix for Andrew.

$$\begin{pmatrix} 3 & 5 & 4 \\ 1 & 4 & 2 \\ 6 & 3 & 7 \end{pmatrix}$$

a Explain why this matrix may be reduced to

$$\begin{pmatrix} 3 & 5 \\ 6 & 3 \end{pmatrix}.$$

b Hence find the best strategy for each player and the value of the game. ⓔ

16 *

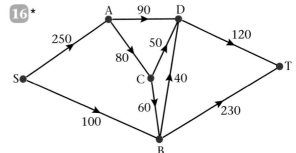

Natural gas is produced at S and is transported to a refinery at T by a network of underwater pipelines. The capacity of each pipeline, in appropriate units, is given in the diagram which shows the network of pipelines

a State the maximum flow along
i SACDT,
ii SBT.

b Show these two maximum flows on Diagram 1 of the worksheet.

c Taking your answer to part **b** as the initial flow pattern, use the labelling procedure to find a maximum flow from S to T showing your working on Diagram 2. List each flow augmenting route you find and state its flow.

d Show your maximum flow pattern on Diagram 3.

e Prove that your flow is maximal.

17 Kris produces custom made racing cycles. She can produce up to four cycles each month, but if she wishes to produce more than three in any one month she has to hire additional help at a cost of £350 for that month. In any month when cycles are produced, the overhead costs are £200. A maximum of three cycles can be held in stock in any one month, at a cost of £40 per cycle per month. Cycles must be delivered at the end of the month. The order book for cycles is

Month	Aug.	Sept.	Oct.	Nov.
Number of cycles required	3	3	5	2

Disregarding the cost of parts and Kris' time,

a determine the total cost of storing two cycles and producing four cycles in a given month, making your calculations clear.

There is no stock at the beginning of August and Kris plans to have no stock after the November delivery.

b Use dynamic programming to determine the production schedule which minimises the costs, showing your working in a table.

The fixed cost of parts is £600 per cycle and of Kris' time is £500 per month. She sells the cycles for £2000 each.

c Determine her total profit for the four-month period. **(E)**

18 *

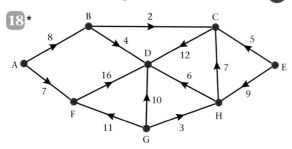

The network above models a drainage system. The number on each arc indicates the capacity of that arc, in litres per second.

a Write down the source vertices.

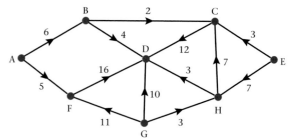

This network shows a feasible flow through the same network.

b State the value of the feasible flow shown.

Taking the flow shown as your initial flow pattern,

c use the labelling procedure to find a maximum flow through this network. You should list each flow-augmenting route you use, together with its flow.

d Show the maximum flow and state its value.

e Prove that your flow is maximal. **(E)**

19 A two-person zero-sum game is represented by the following pay-off matrix for player A.

	B plays 1	B plays 2	B plays 3	B plays 4
A plays 1	−2	1	3	−1
A plays 2	−1	3	2	1
A plays 3	−4	2	0	−1
A plays 4	1	−2	−1	3

a Verify that there is no stable solution to this game.

b Explain why the 4 × 4 game above may be reduced to the following 3 × 3 game.

−2	1	3
−1	3	2
1	−2	−1

c Formulate the 3×3 game as a linear programming problem for player A. Write the constraints as inequalities. Define your variables clearly.

20 a Explain briefly what is meant by a zero-sum game.

A two person zero-sum game is represented by the following pay-off matrix for player A.

	I	II	III
I	5	2	3
II	3	5	4

b Verify that there is no stable solution to this game.

c Find the best strategy for player A and the value of the game to her.

d Formulate the game as a linear programming problem for player B. Write the constraints as inequalities and define your variables clearly.

21 Joan sells ice cream. She needs to decide which three shows to visit over a three-week period in the summer. She starts the three-week period at home and finishes at home. She will spend one week at each of the three shows she chooses, travelling directly from one show to the next.

Table 1 gives the week in which each show is held. Table 2 gives the expected profits from visiting each show. Table 3 gives the cost of travel between shows.

Table 1

Week	1	2	3
Shows	A, B, C	D, E	F, G, H

Table 2

Show	A	B	C	D
Expected profit (£)	900	800	1000	1500

Show	E	F	G	H
Expected profit (£)	1300	500	700	600

Table 3

Travel costs (£)	A	B	C	D
Home	70	80	150	
A				180
B				140
C				200
D				
E				

Travel costs (£)	E	F	G	H
Home		80	90	70
A	150			
B	120			
C	210			
D		200	160	120
E		170	100	110

It is decided to use dynamic programming to find a schedule that maximises the total expected profit, taking into account the travel costs.

a Define suitable stage, state and action variables.

b Determine the schedule that maximises the total profit. Show your working in a table.

c Advise Joan on the shows that she should visit and state her total expected profit.

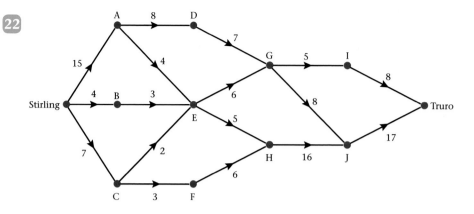

Twenty students wish to travel by bus from Stirling to Truro. The network above shows various routes and the number of free seats available on the coaches that travel on these routes between the two cities. The students agree to travel singly or in small groups and meet up again in Truro.

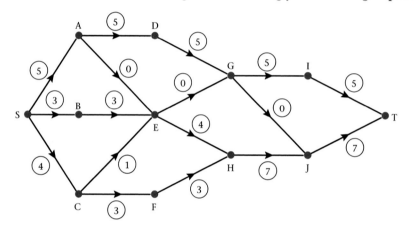

The network above shows how 12 of the students could travel to Truro.

a Using this as your initial flow pattern and listing each flow-augmenting route you use, together with its flow, use the labelling procedure to find the maximum flow.

b Draw a network showing the maximum flow.

c State how many students can travel from Stirling to Truro along these routes.

d Verify your answer using the maximum flow–minimum cut theorem, listing the arcs that your minimum cut passes through. **E**

Examination style paper

There are photocopy masters on the CD-ROM for the questions marked *.

1 At a nursery for a garden centre there are four sites, A, B, C and D available for planting four different types of seedling P, Q, R and S. The numbers of seedlings expected to grow to maturity at each site, and therefore available to be sold in the garden centre, are shown in the table below.

	P	Q	R	S
A	25	29	29	33
B	27	34	33	37
C	29	32	34	37
D	33	36	37	38

Use the Hungarian algorithm, reducing rows first, to obtain an allocation which maximises the total number of seedlings that will grow to maturity. You must make your method clear and show the table after each iteration.

(9)

(Total 9 marks)

2

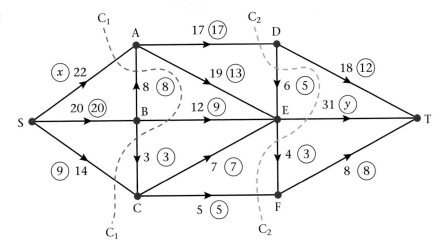

The diagram shows a capacitated, directed network of pipes. The number on each arc represents the capacity of that pipe. The numbers in circles represent a feasible flow.

a State the values of x and y. (2)

b List the saturated arcs. (2)

c State the value of the feasible flow. (1)

d Find the capacities of the cuts C_1 and C_2. (3)

e By inspection, find a flow-augmenting route to increase the flow by 3 units. You must state your route. (1)

f Prove that the new flow is maximal. (2)

(Total 11 marks)

*3

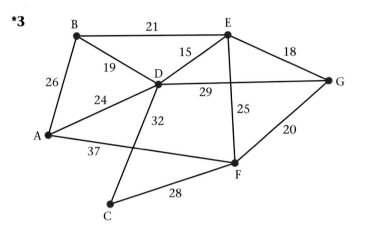

The network in the figure shows the length, in miles, between parcel collection points A, B, C, D, E, F and G. A postal worker will start at A and needs to visit each collection point before returning to A. He wishes to minimise the total distance travelled.

a By inspection complete the table of least distance below. (2)

	A	B	C	D	E	F	G
A	–	26	56	24	☐	37	53
B	26	–	51	19	21	46	39
C	56	51	–	32	☐	28	48
D	24	19	32	–	15	40	29
E	☐	21	☐	15	–	25	18
F	37	46	28	40	25	–	20
G	53	39	48	29	18	20	–

b Use the nearest neighbour algorithm, starting at A, to find an upper bound for the length of the postal worker's route. (2)

c Use the nearest neighbour algorithm, starting at E, to find a second upper bound for the length of the route. (2)

d State the better upper bound of these two, giving a reason for your answer. (1)

e By deleting D find a lower bound for the length of the route. (4)

f Taking your answers to **d** and **e**, use inequalities to write down an interval that must contain the length of the postal worker's optimal route. (1)

(Total 12 marks)

4   The table shows the cost, in pounds, of transporting one unit of stock from each of four supply points A, B, C and D to three demand points P, Q and R. It also shows the stock held at each supply point and the stock required at each demand point.

	P	Q	R	Supply
A	33	26	29	11
B	29	30	32	28
C	31	28	30	36
D	35	32	26	25
Demand	30	40	30	

a Use the north-west corner rule to obtain a possible pattern of distribution   (1)

b Taking the most negative improvement index to indicate the entering square, use the stepping-stone method once to obtain an improved solution. You must make your shadow costs, improvement indices, entering and exiting squares clear.   (6)

c State the cost of your current solution.   (1)

d Determine whether your current solution is optimal, giving a reason for your answer.   (3)

(Total 11 marks)

*5

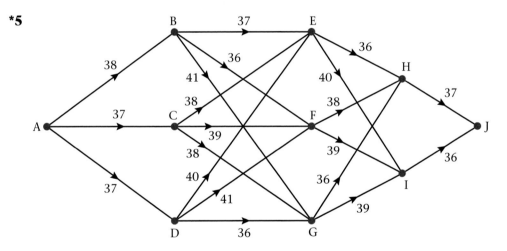

A minimax route is to be found through the network shown in the figure above.
Complete the table on the worksheet, and hence find a minimax route.   (9)

(Total 9 marks)

6   Ben and Greg play a two-person zero-sum game represented by the following pay-off matrix for Ben.

	Greg plays 1	Greg plays 2
Ben plays 1	−3	2
Ben plays 2	2	−1
Ben plays 3	−2	1

Find the best strategy for Greg and the value of the game to him.   (7)

(Total 7 marks)

7* The tableau below is the initial tableau for a maximising linear programming problem in x, y and z.

Basic variable	x	y	z	r	s	t	Value
r	2	1	4	1	0	0	20
s	6	3	2	0	1	0	15
t	1	4	-2	0	0	1	8
P	-3	-1	-4	0	0	0	0

a Taking the most negative number in the profit row to indicate the pivot column at each stage, perform two complete iterations of the simplex algorithm. State the row operations you use and the final values of each variable. (9)

b Determine if your current tableau is optimal, giving a reason for your answer. (1)

(Total 10 marks)

8 A two person zero-sum game is represented by the following pay-off matrix for player A.

	B plays 1	B plays 2	B plays 3
A plays 1	3	1	-2
A plays 2	4	-1	0
A plays 3	0	2	5

Formulate the game as a linear programming problem for player A, writing the constraints as inequalities and stating your variables clearly. (7)

(Total 7 marks)

Answers

Exercise 1A

1 a

	P	Q	R	Supply
A	28	4		32
B		41	3	44
C			34	34
Demand	28	45	37	110

c 22 434

2 a

	P	Q	R	S	Supply
A	21	32	1		54
B			50	17	67
C				29	29
Demand	21	32	51	46	150

c 5032

3 a

	P	Q	R	Supply
A	123			123
B	77	66		143
C		34	50	84
D			150	150
Demand	200	100	200	500

c 8680

4 a

	P	Q	R	S	Supply
A	134				134
B	41	162			203
C		13	163		176
D			12	175	187
Demand	175	175	175	175	700

c 45 761

5 a The total supply is 200, but the total demand is 180. A dummy is needed to absorb this excess, so that total supply equals total demand.

b

	P	Q	R	S	Dummy	Supply
X	40	20				60
Y		50	10			60
Z			40	20	20	80
Demand	40	70	50	20	20	200

5380

6 a A degenerate solution occurs when the number of cells used in a solution is fewer than the number of rows + number of columns − 1. It will happen when an entry, other than the last, completes both the supply requirement of the row and the demand requirement of the column.

b

	K	L	M	N	Stock
A	20				20
B	5	10			15
C			18	2	20
D				20	20
Demand	25	10	18	22	

c

	K	L	M	N	Stock
A	20				20
B	5	10			15
C		0	18	2	20
D				20	20
Demand	25	10	18	22	

or

	K	L	M	N	Stock
A	20				20
B	5	10	0		15
C			18	2	20
D				20	20
Demand	25	10	18	22	

7 a $a = 10$ and $b = 9$

b

	L	M	N	Supply
P	15	7		22
Q		10		10
R			11	11
S			9	9
Demand	15	17	20	

Exercise 1B

1 a

Shadow costs		150	213	227	
		P	Q	R	Supply
0	A	150	213	222	32
−9	B	175	204	218	44
19	C	188	198	246	34
	Demand	28	45	37	110

b Improvement indices for cells:

$BP = 175 + 9 − 150 = 34$
$CP = 188 − 19 − 150 = 19$
$CQ = 198 − 19 − 213 = −34$
$AR = 222 − 0 − 227 = −5$

c Entering cell is CQ, since it has the most negative improvement index.

2 a

Shadow costs		27	33	34	27	
		P	Q	R	S	Supply
0	A	27	33	34	41	54
3	B	31	29	37	30	67
8	C	40	32	28	35	29
	Demand	21	32	51	46	

b Improvement indices for cells:

$BP = 31 − 3 − 27 = 1$
$CP = 40 − 8 − 27 = 5$
$BQ = 29 − 3 − 33 = −7$
$CQ = 32 − 8 − 33 = −9$
$CR = 28 − 8 − 34 = −14$
$AS = 41 − 0 − 27 = 14$

c Entering cell is CR, since it has the most negative improvement index.

3 a

Shadow costs		17	23	19	
		P	Q	R	Supply
0	A	17	24	19	123
−2	B	15	21	25	143
−1	C	19	22	18	84
−3	D	20	27	16	150
	Demand	200	100	200	

b Improvement indices for cells:

$CP = 19 + 1 − 17 = 3$
$DP = 20 + 3 − 17 = 6$
$AQ = 24 − 0 − 23 = 1$
$DQ = 27 + 3 − 23 = 7$
$AR = 19 − 0 − 19 = 0$
$BR = 25 + 2 − 19 = 8$

c There are no negative improvement indices, so the solution is optimal.

4 a

Shadow costs		56	73	60	56	
		P	Q	R	S	Supply
0	A	56	86	80	61	134
3	B	59	76	78	65	203
−3	C	62	70	57	67	176
15	D	60	68	75	71	187
	Demand	175	175	175	175	

b Improvement indices for cells:

$CP = 62 + 3 − 56 = 9$
$DP = 60 − 15 − 56 = −11$
$AQ = 86 − 0 − 73 = 13$
$DQ = 68 − 15 − 73 = −20$
$AR = 80 − 0 − 60 = 20$
$BR = 78 − 3 − 60 = 15$
$AS = 61 − 0 − 56 = 5$
$BS = 65 − 3 − 56 = 6$
$CS = 67 + 3 − 56 = 14$

c Entering cell is DQ, since it has the most negative improvement index.

Exercise 1C

1

		P	Q	R	Supply
	A	28		4	32
	B		11	33	44
	C		34		34
	Demand	28	45	37	150

Cost 21 258

2 21 units A to P
11 units A to Q
22 units A to R
21 units B to Q
46 units B to S
29 units C to R
Cost 4479

3 134 units A to S
163 units B to P
40 units B to S
175 units C to R
1 unit C to S
12 units D to P
175 units D to Q
Cost 43 053

4 1 unit A to P
2 units A to Q
5 units B to P
2 units C to Q
Cost 42

Exercise 1D

1 Let x_{ij} be the number of units transported from i to j where
$i \in \{A, B, C\}$
$j \in \{P, Q, R, S\}$
$x_{ij} \geqslant 0$

Minimise $C = 150x_{11} + 213x_{12} + 222x_{13}$
$+ 175x_{21} + 204x_{22} + 218x_{23}$
$+ 188x_{31} + 198x_{32} + 246x_{33}$

Subject to $x_{11} + x_{12} + x_{13} \leqslant 32$
$x_{21} + x_{22} + x_{23} \leqslant 44$
$x_{31} + x_{32} + x_{33} \leqslant 34$
$x_{11} + x_{21} + x_{31} \leqslant 28$
$x_{12} + x_{22} + x_{32} \leqslant 45$
$x_{13} + x_{23} + x_{33} \leqslant 37$

2 Let x_{ij} be the number of units transported from i to j
where
$i \in \{A, B, C\}$
$j \in \{P, Q, R, S\}$
$x_{ij} \geqslant 0$
Minimise $C = 27x_{11} + 33x_{12} + 34x_{13} + 41x_{14}$
$+ 31x_{21} + 29x_{22} + 37x_{23} + 30x_{24}$
$+ 40x_{31} + 32x_{32} + 28x_{33} + 35x_{34}$

Subject to $x_{11} + x_{12} + x_{13} + x_{14} \leqslant 54$
$x_{21} + x_{22} + x_{23} + x_{24} \leqslant 67$
$x_{31} + x_{32} + x_{33} + x_{34} \leqslant 29$
$x_{11} + x_{21} + x_{31} \leqslant 21$
$x_{12} + x_{22} + x_{32} \leqslant 32$
$x_{13} + x_{23} + x_{33} \leqslant 51$
$x_{14} + x_{24} + x_{34} \leqslant 46$

3 Let x_{ij} be the number of units transported from i to j
where
$i \in \{A, B, C, D\}$
$j \in \{P, Q, R\}$
$x_{ij} \geqslant 0$
Minimise $C = 17x_{11} + 24x_{12} + 19x_{13}$
$+ 15x_{21} + 21x_{22} + 25x_{23}$
$+ 19x_{31} + 22x_{32} + 18x_{33}$
$+ 20x_{41} + 27x_{42} + 16x_{43}$

Subject to $x_{11} + x_{12} + x_{13} \leqslant 123$
$x_{21} + x_{22} + x_{23} \leqslant 143$
$x_{31} + x_{32} + x_{33} \leqslant 84$
$x_{41} + x_{42} + x_{43} \leqslant 150$
$x_{11} + x_{21} + x_{31} + x_{41} \leqslant 200$
$x_{12} + x_{22} + x_{32} + x_{42} \leqslant 100$
$x_{13} + x_{23} + x_{33} + x_{43} \leqslant 200$

4 Let x_{ij} be the number of units transported from i to j
where
$i \in \{A, B, C, D\}$
$j \in \{P, Q, R, S\}$
Minimise $C = 56x_{11} + 86x_{12} + 80x_{13} + 61x_{14}$
$+ 59x_{21} + 76x_{22} + 78x_{23} + 65x_{24}$
$+ 62x_{31} + 70x_{32} + 57x_{33} + 67x_{34}$
$+ 60x_{41} + 68x_{42} + 75x_{43} + 71x_{44}$

Subject to $x_{11} + x_{12} + x_{13} + x_{14} \leqslant 134$
$x_{21} + x_{22} + x_{23} + x_{24} \leqslant 203$
$x_{31} + x_{32} + x_{33} + x_{34} \leqslant 176$
$x_{41} + x_{42} + x_{43} + x_{44} \leqslant 187$
$x_{11} + x_{21} + x_{31} + x_{41} \leqslant 175$
$x_{12} + x_{22} + x_{32} + x_{42} \leqslant 175$
$x_{13} + x_{23} + x_{33} + x_{43} \leqslant 175$
$x_{14} + x_{24} + x_{34} + x_{44} \leqslant 175$

Mixed exercise 1E

1 a

	L	M	Stock
A	15		15
B	1	4	5
C		8	8
Demand	16	12	28

b

Shadow costs		20 L	50 M	Stock
0	A	20	70	15
−20	B	40	30	5
40	C	60	90	8
Demand		16	12	28

c Cost 1140

d Let x_{ij} be the number of units transported from i to j
where
$i \in \{A, B, C\}$
$j \in \{L, M\}$
$x_{ij} \geqslant 0$
Minimise $C = 20x_{11} + 70x_{12}$
$+ 40x_{21} + 30x_{22}$
$+ 60x_{31} + 90x_{32}$

Subject to $x_{11} + x_{12} \leqslant 15$
$x_{21} + x_{22} \leqslant 5$
$x_{31} + x_{32} \leqslant 8$
$x_{11} + x_{21} + x_{31} \leqslant 16$
$x_{12} + x_{22} + x_{32} \leqslant 12$

2 a

	P	Q	R	Stock
F	10	5		15
G		25	10	35
H			10	10
Demand	10	30	20	60

b

Shadow costs		19 P	21 Q	22 R	Stock
0	F	23	21	22	15
2	G	21	23	24	35
0	H	22	21	23	10
Demand		10	30	20	60

c No negative improvement indices

d 1330

e

	P	Q	R	Stock
F			15	15
G	10	20	5	35
H		10		10
Demand	10	30	20	60

Cost 1330

3 **a** Otherwise solution would be degenerate
   b 1675
   c Exiting cell LY cost 1675
   d Not optimal since there are negative improvement indices
   e 25 units M to X
       25 units M to Y
       20 units L to Y
       30 units L to Z
       40 units K to Z
       30 units J to Z
       cost 1135

4 **a** Dummy demand needed to absorb surplus stock

   b

	S	T	U	Dummy	Stock
A	50				50
B	50	20			70
C		10	20	20	50
Demand	100	30	20	20	170

   c Alternative solutions:

	S	T	U	Dummy	Stock
A	30		20		50
B	20	30		20	70
C	50				50
Demand	100	30	20	20	170

	S	T	U	Dummy	Stock
A	50				50
B		30	20	20	70
C	50		0		50
Demand	100	30	20	20	170

	S	T	U	Dummy	Stock
A	50				50
B	0	30	20	20	70
C	50				50
Demand	100	30	20	20	170

   d 910
   e Let x_{ij} be the number of units transported from i to j where
       $i \in \{A, B, C\}$
       $j \in \{S, T, U, \text{dummy}\}$
       $x_{ij} \geq 0$
       Minimise $C = 6x_{11} + 10x_{12} + 7x_{13}$
       $+ 7x_{21} + 5x_{22} + 8x_{23}$
       $+ 6x_{31} + 7x_{32} + 7x_{33}$
       Subject to $x_{11} + x_{12} + x_{13} + x_{14} \leq 50$
       $x_{21} + x_{22} + x_{23} + x_{24} \leq 70$
       $x_{31} + x_{32} + x_{33} + x_{34} \leq 50$
       $x_{11} + x_{21} + x_{31} \leq 100$
       $x_{12} + x_{22} + x_{32} \leq 30$
       $x_{13} + x_{23} + x_{33} \leq 20$
       $x_{14} + x_{24} + x_{34} \leq 20$

Exercise 2A

1 Two solutions

A–Y (35)		A–Z (31)
B–Z (27)	or	B–Y (31)
C–X (30)		C–X (30)
Cost 92		

2 Two solutions

P–A (34)		P–D (32)
Q–B (32)	or	Q–A (35)
R–D (36)		R–B (35)
S–C (35)		S–C (35)
Cost 137		

3 J–R (20)
   K–U (20)
   L–S (10)
   M–T (9)
   Cost 59

4 Two solutions

D–Z (80)		D–Y (87)
E–X (95)		E–X (95)
F–V (90)	or	F–V (90)
G–Y (85)		G–W (83)
H–W (100)		H–Z (95)
Cost 450		

5 Two solutions

A–400 m (64)		A–400 m (64)
B–100 m (13)		B–100 m (13)
C–Hurdles (20)	or	C–200 m (38)
D–200 m (39)		D–Hurdles (21)
Time 136		

6 A–Eucalyptus (62)
   B–Olive (88)
   C–Elm (84)
   D–Beech (145)
   E–Oak (138)
   Cost 517

Exercise 2B

1 J–M (23)
   K–dummy
   L–N (28)
   Cost 51

2 A–Z (35)
   B–Y (10)
   C–W (24)
   dummy–X (0)
   Cost 69

3 Two solutions

W–dummy		W–T (55)
X–T (48)		X–R (67)
Y–S (38)	or	Y–dummy
Z–R (73)		Z–S (37)
Cost 159		

4 P–dummy
   Q–F (39)
   R–G (22)
   S–H (29)
   T–E (18)
   Cost 108

Exercise 2C

1 P–L (48)
Q–M (37)
R–N (56)
Cost 141

2 R–E (47)
S–D (32)
T–F (43)
U–G (47)
Cost 169

3 A–Q (53)
B–R (61)
C–S (62)
D–P (39)
Cost 215

4 J–R (143)
K–S (106)
L–T (143)
M–U (134)
N–V (253)
Cost 779

Exercise 2D

1 L–C (37)
M–E (16)
N–D (41)
Profit 94

2 Two solutions
C–T (34)   C–V (35)
D–U (34) *or* D–U (34)
E–S (42)   E–S (42)
F–V (35)   F–T (34)
Profit 145

3 R–F (22)
S–E (20)
T–G (18)
U–H (28)
Profit 88

4 A–K (95)
B–J (110)
C–N (105)
D–L (84)
E–M (120)
Profit 514

Exercise 2E

1 Let x_{ij} be 0 or 1
$\begin{cases} 1 \text{ if worker } i \text{ does task } j \\ \text{otherwise} \end{cases}$
where $i \in \{L, M, N\}$
    $j \in \{C, D, E\}$
Minimise $C = 37x_{LC} + 15x_{LD} + 12x_{LE}$
    $+ 25x_{MC} + 13x_{MD} + 16x_{ME}$
    $+ 32x_{NC} + 41x_{ND} + 35x_{NE}$
Subject to $\sum x_{Lj} = 1$
    $\sum x_{Mj} = 1$
    $\sum x_{Nj} = 1$
    $\sum x_{iC} = 1$
    $\sum x_{iD} = 1$
    $\sum x_{iE} = 1$

2 Let x_{ij} be 0 or 1
$\begin{cases} 1 \text{ if worker } i \text{ does task } j \\ \text{otherwise} \end{cases}$
where $i \in \{C, D, E, F\}$
    $j \in \{S, T, U, V\}$
Minimise $C = 36x_{CS} + 34x_{CT} + 32x_{CU} + 35x_{CV}$
    $+ 37x_{DS} + 32x_{DT} + 34x_{DU} + 33x_{DV}$
    $+ 42x_{ES} + 35x_{ET} + 37x_{EU} + 36x_{EV}$
    $+ 39x_{FS} + 34x_{FT} + 35x_{FU} + 35x_{FV}$
Subject to $\sum x_{Cj} = 1$ $\sum x_{iS} = 1$
    $\sum x_{Dj} = 1$ $\sum x_{iT} = 1$
    $\sum x_{Ej} = 1$ $\sum x_{iU} = 1$
    $\sum x_{Fj} = 1$ $\sum x_{iV} = 1$

3 Let x_{ij} be 0 or 1
$\begin{cases} 1 \text{ if worker } i \text{ does task } j \\ \text{otherwise} \end{cases}$
where $i \in \{L, M, N\}$
    $j \in \{C, D, E\}$
Minimise $P = 4x_{LC} + 26x_{LD} + 29x_{LE}$
    $+ 16x_{MC} + 28x_{MD} + 25x_{ME}$
    $+ 9x_{NC} +$ $6x_{NE}$
Subject to $\sum x_{Lj} = 1$ $\sum x_{iC} = 1$
    $\sum x_{Mj} = 1$ $\sum x_{iD} = 1$
    $\sum x_{Nj} = 1$ $\sum x_{iE} = 1$

4 Let x_{ij} be 0 or 1
$\begin{cases} 1 \text{ if worker } i \text{ does task } j \\ \text{otherwise} \end{cases}$
where $i \in \{C, D, E, F\}$ and $j \in \{S, T, U, V\}$
Minimise $P = 6x_{CS} + 8x_{CT} + 10x_{CU} + 7x_{CV}$
    $+ 5x_{DS} + 10x_{DT} + 8x_{DU} + 9x_{DV}$
    $+ 7x_{ET} + 5x_{EU} + 6x_{EV}$
    $+ 3x_{FS} + 8x_{FT} + 7x_{FU} + 7x_{FV}$
Subject to $\sum x_{Cj} = 1$ $\sum x_{iS} = 1$
    $\sum x_{Dj} = 1$ $\sum x_{iT} = 1$
    $\sum x_{Ej} = 1$ $\sum x_{iU} = 1$
    $\sum x_{Fj} = 1$ $\sum x_{iV} = 1$

Mixed exercise 2F

1 Bring-it–Depot (326)
Collect-it–Airport (318)
Fetch-it–Docks (317)
Haul-it–Station (321)
Cost 1282

2 a $\begin{pmatrix} 1 & 0 & 0 & 0 \\ 2 & 1 & 0 & 0 \\ 0 & 0 & 1 & 2 \\ 2 & 0 & 0 & 0 \end{pmatrix}$

b and **c** There are 4 solutions each of duration 71
seconds

J–Br (20)   J–C (14)   J–Br (20)   J–Bu (19)
K–Bu (19) *or* K–Bu (14) *or* K–Cr (14) *or* K–Cr (14)
L–Ba (17)   L–Ba (17)   L–Ba (17)   L–Ba (17)
M–C (15)   M–Br (21)   M–Bu (20)   M–Br (21)

3 a Alf–Kitchen (12)
Betty–Gallery (14)
Charlie–Bedroom (14)
Donna–Dining Room (15)
Eve–Grand Hall (14)
Minimum time 69 minutes

b Two solutions

Alf–Dining Room (19)		Alf–Dining Room (19)
Betty–Hall (12)		Betty–Bedroom (18)
Charlie–Gallery (18)	*or*	Charlie–Gallery (18)
Donna–Kitchen (21)		Donna–Kitchen (21)
Eve–Bedroom (20)		Eve–Hall (14)

Maximum time 90 minutes

4 a There are 4 chauffers but only 3 tasks

b D–Party (459)
E–Dummy
F–Film (350)
G–Award (231)
Cost 1040

5 Blue–Computer (636)
Green–Post (674)
Orange–Cleaning (825)
Red–Catering (635)
Yellow–Copying (554)
Cost 3324

6 Ghost train–Coffee shop (365)
Log flume–Cafe (874)
Roller coaster–Snack shop (665)
Teddie's adventure–Restaurant (794)
Profit 2698

7 Let x_{ij} be 0 or 1

$\begin{cases} 1 \text{ if worker } i \text{ does task } j \\ \text{otherwise} \end{cases}$

where $i \in \{P, Q, R, S\}$
$j \in \{1, 2, 3, 4\}$

Minimise $C = 143x_{P1} + 243x_{P2} + 247x_{P3} + 475x_{P4}$
$+ 132x_{Q1} + 238x_{Q2} + 218x_{Q3} + 437x_{Q4}$
$+ 126x_{R1} + 207x_{R2} + 197x_{R3} + 408x_{R4}$
$+ 138x_{S1} + 222x_{S2} + 238x_{S3} + 445x_{S4}$

Subject to $\sum x_{Pj} = 1 \qquad \sum x_{i1} = 1$
$\sum x_{Qj} = 1 \qquad \sum x_{i2} = 1$
$\sum x_{Rj} = 1 \qquad \sum x_{i3} = 1$
$\sum x_{Sj} = 1 \qquad \sum x_{i4} = 1$

8 Let x_{ij} be 0 or 1

$\begin{cases} 1 \text{ if worker } i \text{ does task } j \\ \text{otherwise} \end{cases}$

where $i \in \{P, Q, R, S\}$
$j \in \{A, B, C, D\}$

Minimise $C = 11x_{PA} + 7x_{PB} + 9x_{PC} + 6x_{PD}$
$+ 9x_{QA} + 5x_{QB} + 12x_{QC} + 5x_{QD}$
$+ 8x_{RA} + 4x_{RB} + 11x_{RC} + 2x_{RD}$
$+ 10x_{SA} + 9x_{SB} + 7x_{SC}$

Subject to $\sum x_{Pj} = 1 \qquad \sum x_{iA} = 1$
$\sum x_{Qj} = 1 \qquad \sum x_{iB} = 1$
$\sum x_{Rj} = 1 \qquad \sum x_{iC} = 1$
$\sum x_{Sj} = 1 \qquad \sum x_{iD} = 1$

Exercise 3A

1

	A	B	C	D	E
A	–	7	**10**	9	5
B	7	–	3	**11**	12
C	**10**	3	–	8	12
D	9	**11**	8	–	4
E	5	**12**	12	4	–

AC–the shortest route is ABC length 10
AD–the shortest route is AED length 9
BD–the shortest route is BCD length 11
BE–the shortest route is BAE length 12

2

	A	B	C	D	E
A	–	5	**12**	7	**9**
B	5	–	7	2	**4**
C	**12**	7	–	5	7
D	7	2	5	–	2
E	**9**	**4**	7	2	–

AC–the shortest route is ABDC length 12
AE–the shortest route is ABDE length 9
BC–the shortest route is BDC length 7
BE–the shortest route is BDE length 4
CE–the shortest route is CDE length 7

3

	A	B	C	D	E	F
A	–	10	**18**	13	15	**18**
B	10	–	8	3	**5**	8
C	**18**	8	–	5	3	**10**
D	13	3	**5**	–	2	5
E	15	**5**	3	2	–	**7**
F	**18**	8	**10**	5	**7**	–

AC–the shortest route is ABDEC length 18
AF–the shortest route is ABDF length 18
BC–the shortest route is BDEC length 8
BE–the shortest route is BDE length 5
BF–the shortest route is BDF length 8
CD–the shortest route is CED length 5
CF–the shortest route is CEDF length 10
EF–the shortest route is EDF length 7

4

	A	B	C	D	E	F	G
A	–	10	9	10	17	**20**	**20**
B	10	–	3	20	**11**	10	20
C	9	3	–	19	8	**13**	**23**
D	10	20	19	–	**27**	20	10
E	17	**11**	8	**27**	–	8	18
F	**20**	10	**13**	20	8	–	10
G	**20**	20	**23**	10	18	10	–

AC–the shortest route is ABF length 20
AG–the shortest route is ADG length 20
BE–the shortest route is BCE length 11
CF–the shortest rotue is CBF length 13
CG–the shortest route is CBFG length 23
DE–the shortest route is DACE length 27

Exercise 3B

1 a

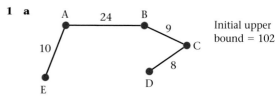

Initial upper bound = 102

b DE shortcut route length 79

c ABCDEA length 79

2 a

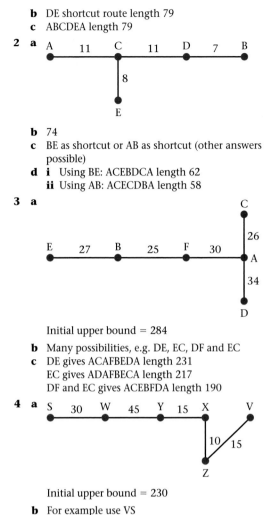

b 74

c BE as shortcut or AB as shortcut (other answers possible)

d **i** Using BE: ACEBDCA length 62

  ii Using AB: ACECDBA length 58

3 a

Initial upper bound = 284

b Many possibilities, e.g. DE, EC, DF and EC

c DE gives ACAFBEDA length 231
EC gives ADAFBECA length 217
DF and EC gives ACEBFDA length 190

4 a

Initial upper bound = 230

b For example use VS

c Route SWYXZVS length 190

Exercise 3C

1 a 45

b This is the route so it is optimal

2 a 50 (deleting A)   49 (deleting B)

b Better lower bound is 50 since it is the highest

3 a 180 (deleting A)   177 (deleting B)

b Better when bound is 180 because it is higher

c $180 <$ optimal value $\leqslant 190$

4 a 170 (deleting S)   145 (deleting V)

b Better lower bound is 170 because it is higher

c $170 <$ optimal value $\leqslant 190$

Exercise 3D

1 a $D_7B_{12}C_8E_{14}A_{19}D_{= 60}$

b $E_8C_{11}A_{13}B_7D_{14}E_{= 53}$ *or* $E_8C_{11}D_7B_{13}A_{14}E_{= 53}$

c The better upper bound is 53 since this is lower

2 a $Z_{10}X_{15}Y_{40}V_{55}W_{30}S_{70}Z_{= 220}$

b $X_{10}Z_{15}V_{40}Y_{45}W_{30}S_{55}X_{= 195}$
$V_{15}Z_{10}X_{15}Y_{45}W_{30}S_{75}V_{= 190}$

c The better upper bound is 190 because it is lower

3 a 1200 minutes

b $U_{120}S_{150}R_{120}V_{150}T_{180}W_{270}U_{= 990}$
and
$U_{120}T_{150}V_{120}R_{150}S_{240}W_{270}U_{= 1050}$

c $V_{120}R_{150}S_{120}U_{120}T_{180}W_{300}V_{= 990}$ and
$V_{120}R_{150}U_{120}S_{210}T_{180}W_{300}V_{= 1080}$ and
$V_{120}R_{150}U_{120}T_{180}W_{240}S_{210}V_{= 1020}$

d The better upper bound is 990 because it is lower

Mixed exercise 3E

1 a Either Kruskal's: EF, DE, CD, BD, AC, EG
or Prim's (e.g.): AC, CD, DE, EF, BD, EG

b 7004

c e.g. use AB and DG
Route ACDEFEGDBA length 6005

2 a

	A	B	C	D	E
A	–	7	13	4	3
B	7	–	17	7	10
C	13	17	–	10	13
D	4	7	10	–	5
E	3	10	13	5	–

b $A_3E_5D_7B_{17}C_{13}A_{= 45}$

c AEDBDCA (BC is not on the original network)

3 a SC SF FA AB CD DE

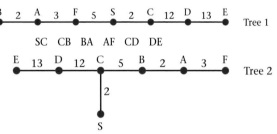

b 74

c Tree 1: use BE as short cut
Route: SCDEBAFS length 56
Tree 2: use EF as short cut
Route: SCBAFEDCS length 53

d $C_2S_5F_3A_2B_{17}D_{13}E_{21}C_{= 63}$
$D_{12}C_2S_5F_3A_2B_{19}E_{13}D_{= 56}$

e The better upper bound is 53 since it is smaller

f Route SCBAFEDCS

g 44

4 a In the classical problem each vertex must be visited *exactly* once before returning to the start.
In the practical problem each vertex must be visited *at least* once before returning to the start.

b

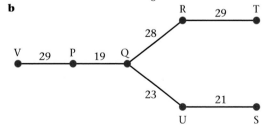

Order: PQ QU US QR $\begin{Bmatrix} TR \\ VP \end{Bmatrix}$

c Use VT and QS as shortcuts giving a length of 213
Route PQUSQRTVP

d $P_{19}Q_{22}U_{21}S_{51}R_{29}T_{37}V_{29}P_{= 209}$

e 186

f $186 <$ optimal value $\leqslant 209$

5 a

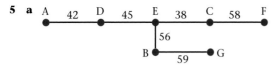

Order: AD DE EC EB CF BG

b 596

c The minimum connector has been doubled and each arc in it repeated

d Use AE and GF as shortcuts
Route: ADEBGFCEA length 427

e 352 km

f The lower bound will give an optimal solution if it is a tour.
If the minimum spanning tree has no 'branches' so the two end vertices have valency 1, and all other vertices have valency 2, then if the two least arcs are incident on the two vertices of valency 1 an optimal solution cannot be found.

6 a

Order: LO OB BN LC OE

b i 824 miles

ii Use NC as a shortcut
Route: LOEOBNCL length 653

c 573

Exercise 4A

1 Let x_1, x_2 and x_3 be the number of round, square and rectangular boxes respectively.

Maximise $P = 12x_1 + 10x_2 + 11x_3$
Subject to: $4x_1 + 2x_2 + 3x_3 + r = 360$
$2x_1 + 3x_2 + 3x_3 + s = 360$
$x_1, x_2, x_3, r, s \geqslant 0$

2 Let x_A, x_B, x_C and x_D be the number of type A, B, C and D backpacks made.

Maximise $P = 8x_A + 7x_B + 6x_C + 9x_D$
Subject to: $2.5x_A + 3x_B + 2x_C + 4x_D + r = 1400$
$10x_A + 12x_B + 8x_C + 15x_D + s = 9000$
$5x_A + 7x_B + 4x_C + 9x_D + t = 4800$
$x_A, x_B, x_C, x_D, r, s, t \geqslant 0$

3 Let x_A, x_C and x_S be the number of adults, children and senior members.

Maximise $P = 40x_A + 10x_C + 20x_S$
Subject to: $x_A + x_C + x_S + r = 100$
$-x_A + x_C - x_S + s = 0$
$-2x_A + x_C + x_S + t = 0$
$x_A, x_C, x_S, r, s, t \geqslant 0$

4 Let x_r, x_f and x_m be the number of batches of rock cakes, fairy cakes and muffins made.

Minimise $T = 10x_r + 20x_f + 15x_m$
Subject to: $220x_r + 100x_f + 250x_m + r = 3000$
$100x_r + 100x_f + 50x_m + s = 2000$
$50x_r + 100x_f + 75x_m + t = 1500$
$8x_r + 18x_f + 12x_m - u = 100$
$x_r, x_f, x_m, r, s, t, u \geqslant 0$

5 Let x_s, x_m and x_l be the number of small, medium and large boxes.

Minimise $C = 0.3x_s + 0.5x_m + 0.8x_l$
Subject to: $x_s + 3x_m + 7x_l - r = 280$
$3x_s + 8x_m + 18x_l - s = 600$
$-x_s + x_m + x_l + t = 0$
$-x_m + 2x_l + u = 0$
$x_s, x_m, x_l, r, s, t, u \geqslant 0$

Exercise 4B

1 $P = 38$ $x = 0$ $y = 3$ $z = 5$ $r = 0$ $s = 0$

2 $P = 260$ $x = 0$ $y = 40$ $z = 10$ $r = 0$ $s = 0$

3 $P = 11$ $x = 2$ $y = 1$ $z = 0$ $r = 0$ $s = 0$ $t = 1$

4 $P = 441$ $x = 105$ $y = 21$ $z = 0$ $r = 441$ $s = 0$ $t = 0$ $u = 79$

5 $P = \frac{877}{8}$ $x_1 = \frac{209}{16}$ $x_2 = \frac{33}{4}$ $x_3 = \frac{261}{16}$ $x_4 = 0$ $r = 0$ $s = 0$ $t = \frac{73}{17}$ $u = 0$

6 *For Q1*

b $P + 12x + r + \frac{4}{3}s = 38$
$\frac{1}{2}x + y + \frac{1}{2}r = 3$
$\frac{7}{6}x + z - \frac{1}{2}r + \frac{1}{3}s = 5$

c $P = 38 - 12x - r - \frac{4}{3}s$
so increasing x, r or s would decrease P.

For Q2

b $P + \frac{1}{2}x + 2r + \frac{3}{2}s = 260$
$\frac{1}{4}x + y + \frac{1}{2}r - \frac{1}{4}s = 40$
$\frac{1}{4}x + z + \frac{1}{4}s = 10$

c $P = 260 - \frac{1}{2}x - 2r - \frac{3}{2}s$, so increasing x, r or s would decrease P.

For Q3

b $P + 8z + \frac{4}{5}r + \frac{3}{5}s = 11$
$x - 5z + \frac{3}{5}r - \frac{4}{5}s = 2$
$y + 5z - \frac{1}{5}r + \frac{3}{5}s = 1$
$15z - r + 2s + t = 1$

c $P = 11 - 8z - \frac{4}{5}r - \frac{3}{5}s$, so increasing z, r or s would decrease P.

For Q4

b $P + 7z + \frac{3}{8}s + \frac{15}{8}t = 441$
$-z + r + \frac{3}{8}s - \frac{17}{8}t = 441$
$x + 7z + \frac{3}{8}s - \frac{1}{8}t = 105$
$y + 3z - \frac{1}{8}s + \frac{3}{8}t = 21$
$9z - \frac{3}{8}s - \frac{7}{8}t + u = 79$

c $P = 441 - 7z - \frac{3}{8}s - \frac{15}{8}t$, so increasing z, s or t would decrease P.

For Q5

b $P + \frac{1}{8}x_4 + \frac{1}{8}r + \frac{1}{4}s + \frac{9}{8}u = \frac{877}{8}$
$x_2 - \frac{3}{4}x_4 + \frac{1}{4}r - \frac{1}{2}s + \frac{1}{4}u = \frac{33}{4}$
$x_3 + \frac{17}{16}x_4 + \frac{1}{16}r + \frac{5}{8}s - \frac{7}{16}u = \frac{261}{16}$
$\frac{21}{16}x_4 - \frac{11}{16}r + \frac{1}{8}s + t - \frac{3}{16}u = \frac{73}{16}$
$x_1 + \frac{13}{16}x_4 - \frac{3}{16}r + \frac{1}{8}s + \frac{5}{16}u = \frac{209}{16}$

c $P = \frac{877}{8} - \frac{1}{8}x_4 - \frac{1}{8}r - \frac{1}{4}s - \frac{9}{8}u$, so increasing x_4, r, s or u would decrease P.

Mixed exercise 4C

1 a There are no negative numbers in the profit row.

b $P + \frac{3}{2}x + \frac{3}{4}r = 840$

so $P = 840 - \frac{3}{2}x - \frac{3}{4}r$

Increasing x or r would decrease P.

c i Maximum profit = £840

ii Optimum number of $A = 0$, $B = 56$ and $C = 75$

2 a Maximise $\quad P = 14x + 20y + 30z$

Subject to: $\quad 5x + 8y + 10z + r = 25\,000$

$\qquad\qquad\quad 5x + 6y + 15z + s = 36\,000$

where r and s are slack variables $x, y, z, r, s \geqslant 0$

b

b.v.	x	y	z	r	s	Value
r	$1\frac{2}{3}$	④	0	1	$-\frac{2}{3}$	1000
z	$\frac{1}{3}$	$\frac{2}{5}$	1	0	$\frac{1}{15}$	2400
P	-4	-8	0	0	2	$72\,000$

b.v.	x	y	z	r	s	Value	Row operation
y	⑤⁄₁₂	1	0	$\frac{1}{4}$	$-\frac{1}{6}$	250	R1 ÷ 4
z	$\frac{1}{6}$	0	1	$-\frac{1}{10}$	$\frac{2}{15}$	2300	R2 $-\frac{2}{5}$R1
P	$-\frac{2}{3}$	0	0	2	$\frac{2}{3}$	$74\,000$	R3 + 8R1

b.v.	x	y	z	r	s	Value	Row operation
x	1	$2\frac{2}{5}$	0	$\frac{3}{5}$	$-\frac{2}{5}$	600	R1 $\div \frac{5}{12}$
z	0	$-\frac{2}{5}$	1	$-\frac{1}{5}$	$\frac{1}{5}$	2200	R2 $-\frac{1}{6}$R1
P	0	$1\frac{3}{5}$	0	$2\frac{2}{5}$	$\frac{2}{5}$	$74\,400$	R3 $+\frac{2}{3}$R1

c i $x = 600 \quad y = 0 \quad z = 2200$

ii Profit is £744

iii The solution is optimal since there are no negative numbers in the profit row.

3 a $\frac{1}{5}(x + y + z) \geqslant y \Rightarrow -x + 4y - z \leqslant 0$

$60x + 100y + 160z \leqslant 2000 \Rightarrow 3x + 5y + 8z \leqslant 100$

$x \geqslant 0 \quad y \geqslant 0 \quad z \geqslant 0$

b $S = 2x + 4y + 6z$

c There are three variables.

f

b.v.	x	y	z	r	t	Value	Row operation
y	$-\frac{5}{37}$	1	0	$\frac{8}{37}$	$\frac{1}{37}$	$2\frac{26}{37}$	R1 $\div 4\frac{5}{8}$
z	$\frac{17}{37}$	0	1	$-\frac{5}{37}$	$\frac{4}{37}$	$10\frac{30}{37}$	R2 $-\frac{5}{8}$R1
S	$\frac{8}{37}$	0	0	$\frac{2}{37}$	$\frac{28}{37}$	$75\frac{25}{37}$	R3 $+\frac{1}{4}$R1

g There are no negative numbers in the objective row.

h 0 small, 2 medium and 11 large tables (seating 74) at a cost of £1960

4 b

b.v.	x	y	r	s	Value
r	5	⑧	1	0	2000
s	3	2	0	1	720
P	-1.5	-1.75	0	0	0

c Optimal solution $\quad x = 125\frac{5}{7} \quad y = 171\frac{3}{7}$

Integer solutions needed, so point testing gives

$x = 126 \quad y = 171$

d The first point is A if y is increased first

$\qquad\qquad\quad$ (D if x is increased first)

The second point is C

5 b $P = 12x + 24y + 20z$

c

b.v.	x	y	z	r	s	Value
r	3	④	2	1	0	100
s	5	3	4	0	1	125
P	-12	-24	-20	0	0	0

e There are no negative numbers in the profit row.

f Type A = 0 $\quad$ Type B = 15 $\quad$ Type C = 20

Profit = £760

6 a Maximise $P = 14x + 12y + 13z$

Subject to:

Carving $\quad 2x + 2.5y + 1.5z \leqslant 8 \Rightarrow 4x + 5y + 3z \leqslant 16$

Sanding $\quad 25x + 20y + 30z \leqslant 120$

$\qquad\qquad\qquad \Rightarrow 5x + 4y + 6z \leqslant 24$

c

b.v.	x	y	z	r	s	Value	Row operation
x	1	$\frac{5}{4}$	$\frac{3}{4}$	$\frac{1}{4}$	0	4	R1 ÷ 4
s	0	$-\frac{9}{4}$	$\frac{9}{4}$	$-\frac{5}{4}$	1	4	R2 − 5R1
P	0	$\frac{11}{2}$	$-\frac{5}{2}$	$\frac{7}{2}$	0	56	R3 + 14R1

d From a zero stock situation we increase the number of lions to 4. We are increasing the profit from 0 to £56.

Review exercise 1

1 A–H $\qquad$ H

$\quad$ B–P $\qquad$ S

$\quad$ C–S $\quad$ *or* $\quad$ I

$\quad$ D–I $\qquad$ P

$\quad$ (both £1077)

2 a x_{11} no. of coaches from A to D

$\quad x_{12}$ no. of coaches from A to E

$\quad x_{13}$ no. of coaches from A to F

$\quad x_{21}$ no. of coaches from B to D

$\quad x_{22}$ no. of coaches from B to E

$\quad x_{23}$ no. of coaches from B to F

$\quad x_{31}$ no. of coaches from C to D

$\quad x_{32}$ no. of coaches from C to E

$\quad x_{33}$ no. of coaches from C to F

b Minimise $\quad z = 40x_{11} + 70x_{12} + 25x_{13}$

$\qquad\qquad\qquad + 20x_{21} + 40x_{22} + 10x_{23}$

$\qquad\qquad\qquad + 35x_{31} + 85x_{32} + 15x_{33}$

c Depot A $\quad x_{11} + x_{12} + x_{13} = 8$ (no. of coaches at A)

$\quad$ Depot B $\quad x_{21} + x_{22} + x_{23} = 5$ (no. of coaches at B)

$\quad$ Depot C $\quad x_{31} + x_{32} + x_{33} = 7$ (no. of coaches at C)

$\quad$ Depot D $\quad x_{11} + x_{21} + x_{31} = 4$ (no. of coaches at D)

$\quad$ Depot E $\quad x_{21} + x_{22} + x_{32} = 10$ (no. of coaches at E)

$\quad$ Depot F $\quad x_{31} + x_{32} + x_{33} = 6$ (no. of coaches at F)

Reference to number of coaches at A, B and C

= number of coaches at D, E and F

3 a

	A	B	C	D	E	F
A	0	20	30	32	12	15
B	20	0	10	(25)	(32)	16
C	30	10	0	15	(35)	19
D	32	(25)	15	0	20	(34)
E	12	(32)	(35)	20	0	16
F	15	16	19	(34)	16	0

b 101 km tour AEFBCDA

c In the original network AD is not a direct path.
The tour becomes AEFBCDEA

d e.g. BCDEAFB
CBFAEDC
DCBFAED $\Big\}$ length 88
EAFBCDE
FAEDCBF

4 a There are no negative entries in the objective row.

b Profit equation $P + z + r + s = 33$
$$P = 33 - (z + r + s)$$
At present z, r and s are all zero. If they increase P
will decrease. Hence P is maximal.

c i $P = 33$
ii $x = 3$ $y = 1$ $z = 0$

5 a

	D	E	F
A	20	4	
B		26	6
C			14

b $S_A = 0$ $S_B = -1$ $S_C = 7$
$D_D = 21$ $D_E = 24$ $D_F = 18$
$I_{AF} = 16 - 0 - 18 = -2$
$I_{BD} = 18 + 1 - 21 = -2$
$I_{CD} = 15 - 7 - 21 = -13^\star$
$I_{CE} = 19 - 7 - 24 = -12$

c

	D	E	F
A	$20 - \theta$	$4 + \theta$	
B		$26 - \theta$	$6 + \theta$
C	θ		$14 - \theta$

entering cell CD
$\theta = 14$
exiting cell CF

	D	E	F
A	6	18	
B		12	20
C	14		

cost £1384

6 a I – C, II – A, III – B, IV – D

b 69 minutes

7 a Yes, there are no negative values in the profit row.

b $P = 63, x = 0, y = 7, z = 0, r = \frac{9}{2}, s = \frac{2}{3}, t = 0$

c 9

8 a In the *practical* T.S.P. each vertex must be visited *at least once*.
In the *classical* T.S.P. each vertex must be visited *exactly once*.

b AB, DF, DE, (reject EF) $\begin{Bmatrix} FG \\ AC \end{Bmatrix}$, EH $\begin{Bmatrix} DC \\ or \\ BE \end{Bmatrix}$

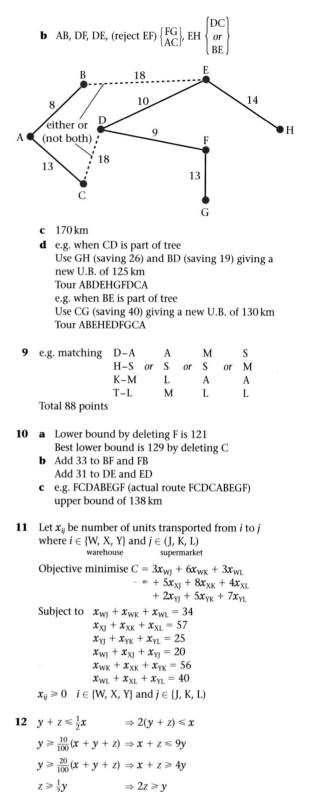

c 170 km

d e.g. when CD is part of tree
Use GH (saving 26) and BD (saving 19) giving a
new U.B. of 125 km
Tour ABDEHGFDCA
e.g. when BE is part of tree
Use CG (saving 40) giving a new U.B. of 130 km
Tour ABEHEDFGCA

9 e.g. matching   D–A   A   M   S
H–S *or* S *or* S *or* M
K–M   L   A   A
T–L   M   L   L

Total 88 points

10 a Lower bound by deleting F is 121
Best lower bound is 129 by deleting C

b Add 33 to BF and FB
Add 31 to DE and ED

c e.g. FCDABEGF (actual route FCDCABEGF)
upper bound of 138 km

11 Let x_{ij} be number of units transported from i to j
where $i \in \{W, X, Y\}$ and $j \in (J, K, L)$
warehouse   supermarket

Objective minimise $C = 3x_{WJ} + 6x_{WK} + 3x_{WL}$
$+ 5x_{XJ} + 8x_{XK} + 4x_{XL}$
$+ 2x_{YJ} + 5x_{YK} + 7x_{YL}$

Subject to $x_{WJ} + x_{WK} + x_{WL} = 34$
$x_{XJ} + x_{XK} + x_{XL} = 57$
$x_{YJ} + x_{YK} + x_{YL} = 25$
$x_{WJ} + x_{XJ} + x_{YJ} = 20$
$x_{WK} + x_{XK} + x_{YK} = 56$
$x_{WL} + x_{XL} + x_{YL} = 40$
$x_{ij} \geqslant 0$ $i \in \{W, X, Y\}$ and $j \in \{J, K, L\}$

12 $y + z \leqslant \frac{1}{2}x$ $\Rightarrow 2(y + z) \leqslant x$
$y \geqslant \frac{10}{100}(x + y + z) \Rightarrow x + z \leqslant 9y$
$y \geqslant \frac{20}{100}(x + y + z) \Rightarrow x + z \geqslant 4y$
$z \geqslant \frac{1}{2}y$ $\Rightarrow 2z \geqslant y$
$x \geqslant 0, y \geqslant 0, z \geqslant 0,$
$x + y + z \geqslant 250$
objective function: minimise; $C = 20x + 26y + 36z$

13 a ii C–III, J–I or IV, N–II, S–IV or I
83 minutes $\therefore$ 11:23 am

b Subtracting all entries from some $n \geqslant 36$
e.g. subtracting from 36

	I	II	III	IV
C	24	2	8	20
J	23	4	0	24
N	21	4	4	22
S	25	3	0	26

14 a i 714   **ii** 552 (ACBDEC)
  b 472   **c** $472 \leqslant$ solution $\leqslant 552$

15 Let x_{ij} be the number of units transported from i to j, in 1000 litres where $i \in \{F, G, H\}$ and $j \in \{S, T, U\}$

Minimise $\quad C = 23x_{FS} + 31x_{FT} + 46x_{FU}$
$\qquad\qquad + 35x_{GS} + 38x_{GT} + 51x_{GU}$
$\qquad\qquad + 41x_{HS} + 50x_{HT} + 63x_{HU}$ unbalanced

Subject to $\quad x_{FS} + x_{FT} + x_{FU} \leqslant 540$
$\qquad\qquad x_{GS} + x_{GT} + x_{GU} \leqslant 789$
$\qquad\qquad x_{HS} + x_{HT} + x_{HU} \leqslant 673$
$\qquad\qquad \left.\begin{array}{l} x_{FS} + x_{GS} + x_{HS} \leqslant 257 \\ x_{FT} + x_{GT} + x_{HT} \leqslant 348 \\ x_{FU} + x_{GU} + x_{HU} \leqslant 410 \end{array}\right\}$ accept = here

$x_{ij} \geqslant 0$

16 a 45
  b i AEFCDBA – length 49
    ii Choose a tour that does not use AB
      e.g. DB(6), BC(10), CF(8), FE(7), EA(7)
      Complete with AD(8), DBCFEAD
      Total length 46

17 a

	W_1	W_2	W_3	Available
F_1	2	2		4
F_2		3		3
F_3		4	4	8
Require	2	9	4	

Cost $2 \times 7 + 2 \times 8 + 3 \times 2 + 4 \times 6 + 4 \times 3$
$= 14 + 16 + 6 + 24 + 12 = 72$

b

Shadow costs		7	8	5	
		W_1	W_2	W_3	
0	F_1	7	8		4
−6	F_2		2		3
−2	F_3		6	3	8
		2	9	4	

$F_1 = 0 \qquad W_1 = 7$
$F_2 = -6 \qquad W_2 = 8$
$F_3 = -2 \qquad W_3 = 5$

c No negative improvement indices and so given solution is optimal and gives minimum cost. If there was a negative I_{ij} then using this route would reduce cost.

18 a If the number of variables $\geqslant 3$ use simplex
  b Column y
  c

b.v.	x	y	z	r	s	Value	Row ops
y	$\frac{9}{14}$	1	0	$\frac{2}{7}$	$-\frac{1}{14}$	$340\frac{4}{7}$	$R1 - \frac{2}{5}R2$
z	$\frac{11}{28}$	0	1	$-\frac{3}{14}$	$\frac{5}{8}$	$222\frac{1}{14}$	$R2 \div 5\frac{3}{5}$
P	$-\frac{2}{7}$	0	0	$1\frac{3}{7}$	$\frac{1}{7}$	$3612\frac{6}{7}$	$R3 + \frac{4}{5}R2$

d $P = 3612\frac{6}{7}$ $x = 0$ $y = 340\frac{4}{7}$ $z = 222\frac{1}{14}$
e No, bottom row still contains a negative. x can be increased.

19 c Machine 1 – Job 2 (5)
    Machine 2 – Job 4 (5)
    Machine 3 – Job 3 (3)
    Machine 4 – Job 1 (2)
    Minimum time: 15 hours

20 a Adds zeros for costs in the third column
    Adds 14 as the demand value
  b The total supply is greater than the total demand
  c The solution would otherwise be degenerate
  d $I_{AJ} = 12 - 0 - 8 = 4$
    $I_{AK} = 15 - 0 - 13 = 2$
    $I_{BK} = 17 - 0 - 13 = 4$
    $I_{CL} = 0 + 4 - 0 = 4$
    No negatives, so optimal

21 a $\therefore$ Best lower bound is 595 km, by deleting C as it is highest lower bound
  b Adds   167 to AF and FA
         137 to CH and HC
         136 to DF and FD
         145 to DG and GD
  c Upper bound, starting at C = 767 km (CDEFHABGC)
    $\therefore$ Best upper bound is 707 starting at F as it is the lowest upper bound found

22 a Idea of many supply and demand points and many units to be moved. Costs are variable and dependant upon the supply and demand points, need to minimise costs. *Practical*
  b Supply = 120   Demand = 110 so not balanced
  c Adds 0, 0, 0, 10 to column f

	d	e	f	
A	45			
B	5	30		
C		30	10	Cost 545

d $R_1 = 0 \qquad R_2 = -1 \qquad R_3 = -3$
    $K_1 = 5 \qquad K_2 = 7 \qquad K_3 = 3$
    $Ae = 3 - 0 - 7 = -4$
    $Af = 0 - 0 - 3 = -3$
    $Bf = 0 + 1 - 3 = -2$
    $Cd = 2 + 3 - 5 = 0$

e

	d	e	f	
A	15	30		
B	35			
C		30	10	Cost 425

23 a A–2   B–4   C–3   D–1   *or*
    A–3   B–4   C–1   D–2
  b £1 160 000
  c Gives other solution not given in **a**

24 a e.g.

	D	E	F
A	6		
B	0	5	
C		4	4

or

	D	E	F
A	6	0	
B		5	
C		4	4

Cost £470

b $S_A = 0$ $S_B = 0$ $S_C = -10$
$S_A = 0$ $S_B = -10$ $S_C = -20$

$D_D = 20$ $D_E = 30$ $D_F = 40$
$D_D = 20$ $D_E = 40$ $D_F = 50$

$I_{AE} = 40 - 30 = 10$ $I_{AF} = 10 - 50 = -40$
$I_{AF} = 10 - 40 = -30$ $I_{BD} = 20 - 10 = 10$
$I_{BF} = 40 - 40 = 0$ $I_{BF} = 40 - 40 = 0$
$I_{CD} = 10 - 10 = 0$ $I_{CD} = 10 - 0 = 10$

c $I_{AE} = 10$, I_{CD} or $I_{BD} = 0$, $I_{BF} = 30$, $I_{CF} = 30$
∴ optimal cost £350

25 a i Slack variables, used to enable us to write inequalities as equalities
ii Represents the equation $P - 3x - 6y - 4z = 0$
b $P = 15$, $x = 4$, $y = \frac{1}{2}$, $z = 0$, $r = 0$, $s = 0$, $t = 7\frac{1}{2}$
c Profit = £15 when $x = 4\,$kg, $y = \frac{1}{2}\,$kg, $z = 0$
The first and second constraints have no slack.
The third constraint has a slack of $7\frac{1}{2}$

26 a Order of arcs: AB, BC, CF, FD, FE

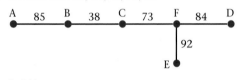

b i 744
ii e.g. AD saves 105 giving 639
or   AE saves 180 giving 564
AF saves 96 giving 648
DE saves 66 giving 678
c 498

27 a Maximise $P = 50x + 80y + 60z$
Subject to $x + y + 2z \leqslant 30$
$x + 2y + z \leqslant 40$
$3x + 2y + z \leqslant 50$
where $x, y, z \geqslant 0$
c The solution found after one iteration has a slack of 10 units of black per day.
d i

b.v.	x	y	z	r	s	t	Value	
z	$\frac{1}{3}$	0	1	$\frac{2}{3}$	$-\frac{1}{3}$	0	$6\frac{2}{3}$	R1 ÷ $\frac{3}{2}$
y	$\frac{1}{3}$	1	0	$-\frac{1}{3}$	$\frac{2}{3}$	0	$16\frac{2}{3}$	R2 − $\frac{1}{2}$R1
t	2	0	0	0	-1	1	10	R3 − no change
P	$-3\frac{1}{3}$	0	0	$13\frac{1}{3}$	$33\frac{1}{3}$	0	$1733\frac{1}{3}$	R4 + 20R1

ii Not optimal, a negative value in profit row
iii $x = 0$ $y = 16\frac{2}{3}$ $z = 6\frac{2}{3}$
$P = £1733.33$ $r = 0$ $s = 0$ $t = 10$

28 a

F — D — A — C — E — B
48   54   53   58   38

b i 502
ii Finding a shortcut to below 360, e.g. FB leaves 351

c M.S.T. is DF, CE, EB, FB length 244
The 2 shortest arcs are AC (53) and AD (54) giving a total of 351
d The optimal solution is 351 and is
A – C – E – B – F – D – A
e

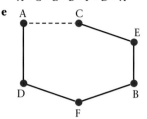

29 a r, s and t are unused amounts of birdseed (in kg), suet blocks and peanuts (in kg) that Polly has at the end of each week after she has made up and sold her packs.
b

b.v.	x	y	z	r	s	t	Value	
z	$\frac{2}{5}$	$\frac{1}{2}$	1	$\frac{1}{10}$	0	0	14	R1 ÷ 10
s	$\boxed{\frac{2}{5}}$	-1	0	$-\frac{2}{5}$	1	0	4	R2 − 4R1
t	$-\frac{1}{5}$	$\frac{1}{2}$	0	$-\frac{3}{10}$	0	1	18	R3 − 3R3
P	-90	-25	0	65	0	0	9100	R4 + 650R1

c $x = 0$ $y = 0$ $z = 14$ $r = 0$ $s = 4$
$t = 18$ $P = £91$
d $P = 90x - 25y + 65r = 9100$
e $P = 9100 + 90x + 25y - 65r$
So increasing x or y would increase the profit.
f The $\frac{2}{5}$ in the x column and 2nd (s) row.

30 a

	B_1	B_2	B_3
F_1	20	15	
F_2		10	15
F_1			15

b $I_{13} = 11 - 0 - 7 = 4$
$I_{21} = 12 - 1 - 10 = 1$
$I_{31} = 9 - 0 - 10 = -1$
$I_{32} = 6 - 0 - 4 = 2$
Since I_{31} is negative, pattern is not optimal.

c

	B_1	B_2	B_3
F_1	10	25	
F_2			25
F_1	10		5

d cost 525 units

31 a $x + 2y + 4z \leqslant 24$
b i $x + 2y + 4z + s = 24$
ii $s(\geqslant 0)$ is the slack time on the machine in hours.
c 1 euro
d Profit = 31 euros $y = 7$ $z = 2.5$ $x = r = s = 0$
e Cannot make $\frac{1}{2}$ a lamp
f e.g. (0, 10, 0) or (0, 6, 3) or (1, 7, 2)
checks in **both** inequalities

32 a

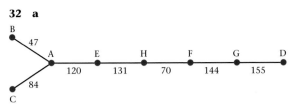

b 1502

c B to D saves 265, H to G saves 54, B to H saves 133 etc.

d 971

e The non-deleted vertices form a minimum spanning tree so they do not form a cycle.
The optimum solution is a cycle.
Unless the 2 least paths complete the cycle it will not give the optimum solution.
In general this will not be the case so a lower bound will be formed, shorter than the optimum solution.

33 b $P = 10x + 20y + 28z$

c

b.v.	x	y	z	r	s	Value
r	1	4	6	1	0	120
s	1	2	5	0	1	100
P	-10	-20	-28	0	0	0

e This tableau is optimal as there are no negative numbers in the profit line.

f small 80, medium 10, large 0, profit £1000

34 a

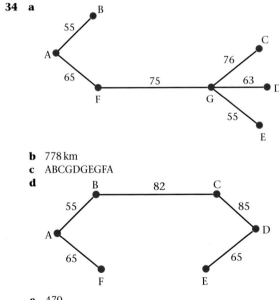

b 778 km

c ABCGDGEGFA

d

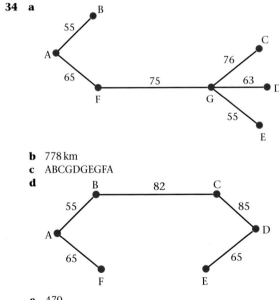

e 470

f Route ABCDEGFA, length 482 km

35 a Objective: Maximise $P = 4x + 5y + 3z$
Subject to $3x + 2y + 4z \leq 35$
$x + 3y + 2z \leq 20$
$2x + 4y + 3z \leq 24$

b $P = 47\frac{1}{4}, x = 11\frac{1}{2}, y = \frac{1}{4}, z = 0, r = 0, s = 7\frac{3}{4}, t = 0$

c There is some slack $(7\frac{3}{4})$ on S, so *do not* increase blending; therefore increase processing and packing which are both at their limit at present.

Exercise 5A

1 a A should play 1 (row maximin = 2)
B should play 2 (column minimax = 2)

b 2 = 2 so game stable

2 a R should play 3 (row maximin = -1)
S should play 3 (column minimax = 1)

b $-1 \neq 1$ so game not stable

3 a A should play 2 or 4 (row maximin = -1)
B should play 2 (column minimax = -1)

b $-1 = -1$ so game stable saddle points (A2, B2) and (A4, B2)

c Value of the game is -1 to A. (If A plays 2 or 4 and B plays 2 the value of the game is -1.)

4 a C plays 2 (row maximin = -1)
D plays 3 (column minimax = 2)

b $-1 \neq 2$ so game not stable

c If C plays 2 and D plays 3, the value of the game is 1 to Claire.

d *Either* Since the value of the game is 1 to Claire and it is a zero-sum game, the value of the game must be -1 to David.
Or If C plays 2 and D plays 3 Claire wins 1, so David wins -1

e

	C plays 1	C plays 2	C plays 3	C plays 4
D plays 1	-7	-4	2	-3
D plays 2	-2	1	-5	3
D plays 3	3	-1	-2	4
D plays 4	-5	-3	1	-2

5 a H plays 1 or 2
D plays 3 or 4

b Saddlepoints (H1, D3) (H2, D3) (H1, D4) (H2, D4)

c The value of the game to Hilary = 0

d The value of the game to Denis = 0

e

	H plays 1	H plays 2	H plays 3	H plays 4	H plays 5
D plays 1	-2	-4	-1	-1	0
D plays 2	-1	0	-4	-1	2
D plays 3	0	0	1	1	3
D plays 4	0	6	1	2	3
D plays 5	-2	-2	-3	0	1

Exercise 5B

1

	Freya plays 1	Freya plays 2
Ellie plays 2	-1	6
Ellie plays 3	3	-3

2

	Harry plays 1	Harry plays 3
Doug plays 1	-5	-1
Doug plays 2	2	-6

3

	Nick plays 1	Nick plays 2
Chris plays 1	1	2
Chris plays 2	2	-1

4 b A should play 1 with probability $\frac{2}{5}$
A should play 2 with probability $\frac{3}{5}$
The value of the game to A is $3(\frac{2}{5}) - 1 = \frac{1}{5}$

c B should play 1 with probability $\frac{7}{10}$

B should play 2 with probability $\frac{3}{10}$

The value of the game to B is $4(\frac{3}{10}) - 3 = -\frac{1}{5}$

5 b A should play 1 with probability $\frac{3}{7}$

A should play 2 with probability $\frac{4}{7}$

The value of the game to A is $2 - 5(\frac{3}{7}) = -\frac{1}{7}$

c B should play 1 with probability $\frac{9}{14}$

B should play 2 with probability $\frac{5}{14}$

The value of the game to B is $8(\frac{9}{14}) - 5 = \frac{1}{7}$

6 b A should play 1 with probability $\frac{1}{3}$

A should play 2 with probability $\frac{2}{3}$

The value of the game to A is $7(\frac{1}{3}) - 2 = \frac{1}{3}$

c B should play 1 with probability $\frac{2}{9}$

B should play 2 with probability $\frac{7}{9}$

The value of the game to B is $1 - 6(\frac{2}{9}) = -\frac{1}{3}$

7 b A should play 1 with probability $\frac{3}{7}$

A should play 2 with probability $\frac{4}{7}$

The value of the game to A is $1 - 2(\frac{3}{7}) = \frac{1}{7}$

c B should play 1 with probability $\frac{5}{7}$

B should play 2 with probability $\frac{2}{7}$

The value of the game to B is $4(\frac{5}{7}) - 3 = -\frac{1}{7}$

Exercise 5C

1 b A should play 1 with probability $\frac{5}{12}$

A should play 2 with probability $\frac{7}{12}$

The value of the game to A is $1 - 6(\frac{5}{12}) = -\frac{3}{2}$

2 b A should play 1 with probability $\frac{1}{2}$

A should play 2 with probability $\frac{1}{2}$

The value of the game to A is $3(\frac{1}{2}) - 1 = \frac{1}{2}$

3 b A should play 1 with probability $\frac{9}{17}$

A should play 2 with probability $\frac{8}{17}$

The value of the game to A is $10(\frac{9}{17}) - 4 = \frac{22}{17}$

4 b A should play 1 with probability $\frac{4}{11}$

A should play 2 with probability $\frac{7}{11}$

The value of the game to A is $1 - 3(\frac{4}{11}) = -\frac{1}{11}$

5 b B should play 1 with probability $\frac{5}{9}$

B should play 2 with probability $\frac{4}{9}$

The value of the game to B is $2(\frac{5}{9}) - 1 = \frac{1}{9}$

6 b B should play 1 with probability $\frac{7}{15}$

B should play 2 with probability $\frac{8}{15}$

The value of the game to B is $9(\frac{7}{15}) - 4 = \frac{3}{15}$

7 b B should play 1 with probability $\frac{6}{11}$

B should play 2 with probability $\frac{5}{11}$

The value of the game to B is $5(\frac{6}{11}) - 2 = \frac{8}{11}$

8 b B should play 1 with probability $\frac{5}{8}$

B should play 2 with probability $\frac{3}{8}$

The value of the game to B is $6(\frac{5}{8}) - 4 = -\frac{1}{4}$

Exercise 5D

1 Let A play 1 with probability p_1

A play 2 with probability p_2

A play 3 with probability p_3

Let the value of the game to A be v and $V = v + 5$

Maximise $P = V$

Subject to $4p_1 + 8p_2 + 3p_3 \geqslant V$

$\Rightarrow V - 4p_1 - 8p_2 - 3p_3 + r = 0$

$6p_1 + p_2 + 7p_3 \geqslant V$

$\Rightarrow V - 6p_1 - p_2 - 7p_3 + s = 0$

$p_1 + p_2 + p_3 \leqslant 1$

$\Rightarrow p_1 + p_2 + p_3 + t = 1$

$p_1, p_2, p_3, r, s, t \geqslant 0$

2 Let A play 1 with probability p_1

A play 2 with probability p_2

A play 3 with probability p_3

Let the value of the game to A be v and $V = v + 6$

Maximise $P = V$

Subject to $p_1 + 9p_2 + 7p_3 \geqslant V$

$\Rightarrow V - p_1 - 9p_2 - 7p_3 + r = 0$

$10p_1 + 3p_2 + 4p_3 \geqslant V$

$\Rightarrow V - 10p_1 - 3p_2 - 4p_3 + s = 0$

$7p_1 + 8p_2 + 5p_3 \geqslant V$

$\Rightarrow V - 7p_1 - 8p_2 - 5p_3 + k = 0$

$p_1 + p_2 + p_3 \leqslant 1$

$\Rightarrow p_1 + p_2 + p_3 + u = 1$

$p_1, p_2, p_3, r, s, t, u \geqslant 0$

3 Let A play 1 with probability p_1

A play 2 with probability p_2

A play 3 with probability p_3

Let the value of the game to A be v and $V = v + 5$

Maximise $P = V$

Subject to $2p_1 + 4p_2 + 7p_3 \geqslant V$

$\Rightarrow V - 2p_1 - 4p_2 - 7p_3 + r = 0$

$7p_1 + 3p_2 + p_3 \geqslant V$

$\Rightarrow V - 7p_1 - 3p_2 - p_3 + s = 0$

$4p_1 + 6p_2 + 3p_3 \geqslant V$

$\Rightarrow V - 4p_1 - 6p_2 - 3p_3 + t = 0$

$p_1 + p_2 + p_3 \leqslant 1$

$\Rightarrow p_1 + p_2 + p_3 + u = 1$

$p_1, p_2, p_3, r, s, t, u \geqslant 0$

4 Let A play 1 with probability p_1

A play 2 with probability p_2

A play 3 with probability p_3

Let the value of the game to A be v and $V = v + 4$

Maximise $P = V$

Subject to $6p_1 + 2p_2 + 5p_3 \geqslant V$

$\Rightarrow V - 6p_1 - 2p_2 - 5p_3 + r = 0$

$p_1 + 8p_2 + 3p_3 \geqslant V$

$\Rightarrow V - p_1 - 8p_2 - 3p_3 + s = 0$

$3p_1 + 5p_2 + 4p_3 \geqslant V$

$\Rightarrow V - 3p_1 - 5p_2 - 4p_3 + t = 0$

$p_1 + p_2 + p_3 \leqslant 1$

$\Rightarrow p_1 + p_2 + p_3 + u = 1$

$p_1, p_2, p_3, r, s, t, u \geqslant 0$

5 Let B play 1 with probability q_1

B play 2 with probability q_2

B play 3 with probability q_3

Let the value of the game to B be v and $V = v + 4$

Maximise $P = V$

Subject to $9q_1 + 2q_2 + q_3 \geq V$
$\Rightarrow V - 9q_1 - 2q_2 - q_3 + r = 0$
$3q_1 + 7q_2 + 8q_3 \geq V$
$\Rightarrow V - 3q_1 - 7q_2 - 8q_3 + s = 0$
$q_1 + q_2 + q_3 \leq 1$
$\Rightarrow q_1 + q_2 + q_3 + t = 1$
$q_1, q_2, q_3, r, s, t \geq 0$

6 Let B play 1 with probability q_1
B play 2 with probability q_2
B play 3 with probability q_3
Let the value of the game to B be v and $V = v + 5$

Maximise $P = V$

Subject to $10q_1 + q_2 + 4q_3 \geq V$
$\Rightarrow V - 10q_1 - q_2 - 4q_3 + r = 0$
$2q_1 + 8q_2 + 3q_3 \geq V$
$\Rightarrow V - 2q_1 - 8q_2 - 3q_3 + s = 0$
$4q_1 + 7q_2 + 6q_3 \geq V$
$\Rightarrow V - 4q_1 - 7q_2 - 6q_3 + t = 0$
$q_1 + q_2 + q_3 \leq 1$
$\Rightarrow q_1 + q_2 + q_3 + u = 1$
$q_1, q_2, q_3, r, s, t, u \geq 0$

7 Let B play 1 with probability q_1
B play 2 with probability q_2
B play 3 with probability q_3
Let the value of the game to B be v and $V = v + 3$

Maximise $P = V$

Subject to $6q_1 + q_2 + 4q_3 \geq V$
$\Rightarrow V - 6q_1 - q_2 - 4q_3 + r = 0$
$4q_1 + 5q_2 + 2q_3 \geq V$
$\Rightarrow V - 4q_1 - 5q_2 - 2q_3 + s = 0$
$q_1 + 7q_2 + 5q_3 \geq V$
$\Rightarrow V - q_1 - 7q_2 - 5q_3 + t = 0$
$q_1 + q_2 + q_3 \leq 1$
$\Rightarrow q_1 + q_2 + q_3 + u = 1$
$q_1, q_2, q_3, r, s, t, u \geq 0$

8 Let B play 1 with probability q_1
B play 2 with probability q_2
B play 3 with probability q_3
Let the value of the game to B be v and $V = v + 5$

Maximise $P = V$

Subject to $3q_1 + 8q_2 + 6q_3 \geq V$
$\Rightarrow V - 3q_1 - 8q_2 - 6q_3 + r = 0$
$7q_1 + q_2 + 4q_3 \geq V$
$\Rightarrow V - 7q_1 - q_2 - 4q_3 + s = 0$
$4q_1 + 6q_2 + 5q_3 \geq V$
$\Rightarrow V - 4q_1 - 6q_2 - 5q_3 + t = 0$
$q_1 + q_2 + q_3 \leq 1$
$\Rightarrow q_1 + q_2 + q_3 + u = 1$
$q_1, q_2, q_3, r, s, t, u \geq 0$

9 a

b.v.	v	p_1	p_2	p_3	r	s	t	Value
r	①	-4	-8	-3	1	0	0	0
s	1	-6	-1	-7	0	1	0	0
t	0	1	1	1	0	0	1	1
P	-1	0	0	0	0	0	0	0

Value of game $= \frac{44}{9} - 5 = -\frac{1}{9}$

b A should play 1 with probability $\frac{7}{9}$, play 2 with probability $\frac{2}{9}$ and play 3 never.

10 a

b.v.	v	q_1	q_2	q_3	r	s	t	Value
r	①	-9	-2	-1	1	0	0	0
s	1	-3	-7	-8	0	1	0	0
t	0	1	1	1	0	0	1	1
P	-1	0	0	0	0	0	0	0

b B should play 1 with probability $\frac{7}{13}$, play 2 never and play 3 with probability $\frac{6}{13}$

Value of game $= \frac{69}{13} - 4 = \frac{17}{13}$

11 a

b.v.	v	p_1	p_2	p_3	r	s	t	u	Value
r	①	-1	-9	-7	1	0	0	0	0
s	1	-10	-3	-4	0	1	0	0	0
t	1	-7	-8	-5	0	0	1	0	0
u	0	1	1	1	0	0	0	1	1
P	-1	0	0	0	0	0	0	0	0

b A should play 1 with probability $\frac{2}{5}$
A should play 2 with probability $\frac{3}{5}$
A should play 3 never
Value of game $= \frac{29}{5} - 6 = -\frac{1}{5}$

c

b.v.	v	q_1	q_2	q_3	r	s	t	u	Value
r	①	-10	-1	-4	1	0	0	0	0
s	1	-2	-8	-3	0	1	0	0	0
t	1	-4	-7	-6	0	0	1	0	0
u	0	1	1	1	0	0	0	1	1
P	-1	0	0	0	0	0	0	0	0

d B should play 1 with probability $\frac{7}{15}$
B should play 2 with probability $\frac{8}{15}$
B should play 3 never
Value of game $= \frac{26}{5} - 5 = \frac{1}{5}$

12 a

b.v.	v	p_1	p_2	p_3	r	s	t	u	Value
r	①	-2	-4	-7	1	0	0	0	0
s	1	-7	-3	-1	0	1	0	0	0
t	1	-4	-6	-3	0	0	1	0	0
u	0	1	1	1	0	0	0	1	1
P	-1	0	0	0	0	0	0	0	0

b A should play 1 with probability $\frac{4}{9}$
A should play 2 with probability $\frac{2}{9}$
A should play 3 with probability $\frac{3}{9}$
Value of game $= \frac{37}{9} - 5 = -\frac{8}{9}$

c

b.v.	v	q_1	q_2	q_3	r	s	t	u	Value
r	①	-6	-1	-4	1	0	0	0	0
s	1	-4	-5	-2	0	1	0	0	0
t	1	-1	-7	-5	0	0	1	0	0
u	0	1	1	1	0	0	0	1	1
P	-1	0	0	0	0	0	0	0	0

d B should play 1 with probability $\frac{4}{9}$

B should play 2 with probability $\frac{3}{9}$

B should play 3 with probability $\frac{2}{9}$

Value of game $= \frac{35}{9} - 3 = \frac{8}{9}$

Mixed exercise 5E

1 A should play 1 with probability $\frac{11}{17}$

A should play 2 with probability $\frac{6}{17}$

The value of the game to A is $\frac{14}{17}$

B should play 1 with probability $\frac{8}{17}$

B should play 2 with probability $\frac{9}{17}$

The value of the game to B is $-\frac{14}{17}$

2 a column 3 dominates column 2 (since $3 < 4$ and $-4 < -1$)

b A should play 1 with probability $\frac{5}{13}$

A should play 2 with probability $\frac{8}{13}$

The value of the game is $-\frac{17}{13}$

B should play 1 with probability $\frac{7}{13}$

B should play 2 with probability $\frac{6}{13}$

The value of the game is $\frac{17}{13}$

3 c Row 1 dominates row 3

(since $-5 > -7$ $2 > 0$ $3 > 1$)

	G plays 1	G plays 2	G plays 3
C plays 1	-5	2	3
C plays 2	1	-3	-4

d Cait should play 1 with probability $\frac{5}{13}$

Cait should play 2 with probability $\frac{8}{13}$

Cait should play 3 never

The value of the game is $-\frac{17}{13}$

e The value of the game to Georgi is $\frac{17}{13}$

4 b A row x dominates a row y if in each column the element in row $x \geqslant$ the element in row y.

c Row 4 dominates row 3

	B plays 1	B plays 2	B plays 3
A plays 1	2	-1	-3
A plays 2	-2	1	4
A plays 3	-1	2	-2

d Let A play 1 with probability p_1

Let A play 2 with probability p_2

Let A play 3 with probability p_3

Let the value of the game to A be v so $V = v + 4$

Maximise $P = V$

Subject to: $6p_1 + 2p_2 + 3p_3 \geqslant V$

$3p_1 + 5p_2 + 6p_3 \geqslant V$

$p_1 + 8p_2 + 2p_3 \geqslant V$

$p_1 + p_2 + p_3 \leqslant 1$

5 a A plays 1, B plays 2

c

	B plays 2	B plays 3
A plays 1	-3	1
A plays 2	-4	4
A plays 3	2	-1

d

	A plays 1	A plays 2	A plays 3
B plays 2	3	4	-2
B plays 3	-1	-4	1

e B should play 2 with probability $\frac{5}{11}$

B should play 3 with probability $\frac{6}{11}$

The value of the game is $-\frac{4}{11}$

6 a A plays 2, B plays 1

c

	A plays 1	A plays 2	A plays 3
B plays 1	-2	-5	2
B plays 2	-7	0	-3
B plays 3	1	-8	-5

d Let B play 1 with probability p_1, play 2 with probability p_2 and play 3 with probability p_3

Let $v =$ value of the game to B so $V = v + 9$

Maximise $P = V$

Subject to: $7p_1 + 2p_2 + 10p_3 \geqslant V$

$\Rightarrow V - 7p_1 - 2p_2 - 10p_3 + r = 0$

$4p_1 + 9p_2 + p_3 \geqslant V$

$\Rightarrow V - 4p_1 - 9p_2 - p_3 + s = 0$

$11p_1 + 6p_2 + 4p_3 \geqslant V$

$\Rightarrow V - 11p_1 - 6p_2 - 4p_3 + t = 0$

$p_1 + p_2 + p_3 \leqslant 1$

$\Rightarrow p_1 + p_2 + p_3 + u = 0$

where $p_1, p_2, p_3, r, s, t, u \geqslant 0$

e

b.v.	v	p_1	p_2	p_3	r	s	t	u	Value
r	1	-7	-2	-10	1	0	0	0	0
s	1	-4	-9	-1	0	1	0	0	0
t	1	-11	-6	-4	0	0	1	0	0
u	0	1	1	1	0	0	0	1	1
P	-1	0	0	0	0	0	0	0	0

Exercise 6A

1 a Flow into B = flow out of B $w = 3$

Flow into A = flow out of A $x = 4$

Flow into E = flow out of E $y = 4$

Flow into D = flow out of D $z = 13$

b Feasible flow = 28

c CE and ED are saturated

d BD has capacity 8

e Along SAT the current flow is 8

2 a Flow into A = flow out of A $w = 9$

Flow into E = flow out of E $x = 5$

Flow into C = flow out of C $y = 2$

Flow into D = flow out of D $14 = y + x + 2$

$\Rightarrow 14 = 2 + 5 + z$

$\Rightarrow z = 7$

b Feasible flow = 38

c BE and AC are saturated

d Flow along SD is 14

e Flow along SBET = 15

3 a Source vertex is F

b Sink vertex is C

c Flow into A = flow out of A $w = 8$

Flow into B = flow out of B $x = 3$

Flow into D = flow out of D $y = 20$

Flow into G = flow out of G $z = 4$

d Feasible flow = 27
e Saturated arcs are AC, FC, FG
f Capacity of FB is 8

4 **a** Source vertex is E
b Sink vertex is C
c Flow into A = flow out of A $\quad w = 5$
Flow into B = flow out of B $\quad x = 3$
Flow into G = flow out of G $\quad y = 4$
Flow into D = flow out of D $\quad z = 5$
d Feasible flow = 20
e Saturated arcs are BA, ED, DG, GF
f Flow along FC = 11

5 e.g.

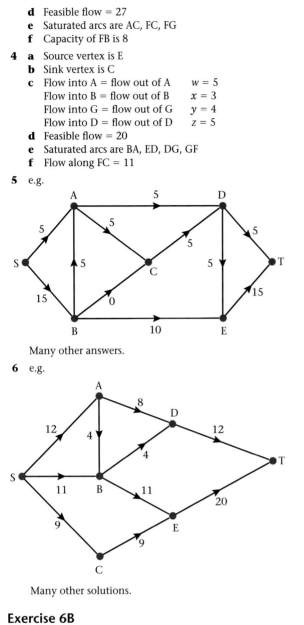

Many other answers.

6 e.g.

Many other solutions.

Exercise 6B

1 Cut C_1 = 19 + 8 + 10 + 4 = 41
Cut C_2 = 20 + 7 + 20 = 47
2 Cut C_3 = 7 + 9 + 4 + 19 = 39
Cut C_4 = 15 + 3 + 16 = 34
3 Cut C_5 = 15 + 45 + 18 + 15 + 10 = 103
Cut C_6 = 15 + 10 + 20 + 10 + 15 + 8 = 78
Cut C_7 = 20 + 45 + 18 + 15 + 8 = 106
4 Cut C_8 = 14 + 14 + 4 + 14 = 50
Cut C_9 = 24 + 14 + 4 + 18 = 60
5 Cut C_{10} = 16 + 16 + 4 + 25 = 61
Cut C_{11} = 30 + 6 + 23 + 10 = 69
6 Cut C_{12} = 30 + 32 + 18 + 30 = 110
Cut C_{13} = 20 + 50 + 18 + 35 = 123
Cut C_{14} = 20 + 15 + 8 + 15 + 10 + 18 + 14 = 100

Exercise 6C

1 **a** max flow along SACT = 13
max flow along SBCDT = 8

b

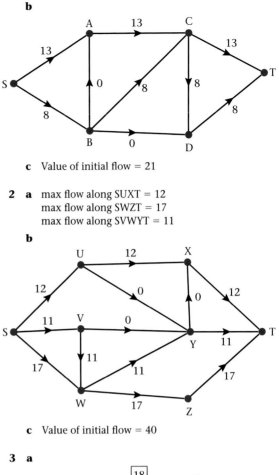

c Value of initial flow = 21

2 **a** max flow along SUXT = 12
max flow along SWZT = 17
max flow along SVWYT = 11

b

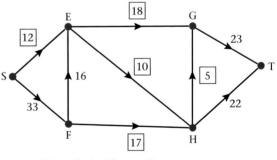

c Value of initial flow = 40

3 **a**

c Value of initial flow = 45

4 **a**

c Value of initial flow = 57

Exercise 6D

1 a

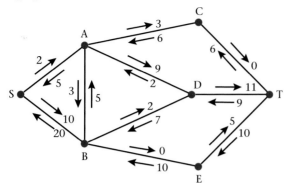

e.g. – there are many other combinations of flows possible

SBADT – 5
SADT – 2
SBDT – 2

b

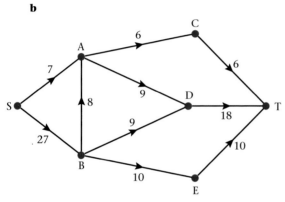

Value of maximum flow is 34

2 a

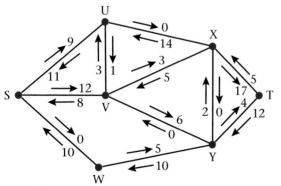

e.g. – there are many other combinations of flows possible

SVYT – 4
SVXT – 3
SVYXT – 2

b

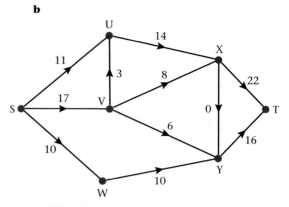

Value of maximum flow is 38

3 a

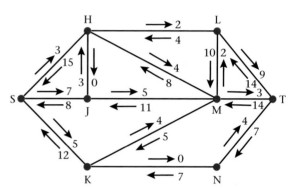

e.g. – there are many other combinations of flows possible

SHMT – 3
SJMLT – 2
SJHLT – 2

b

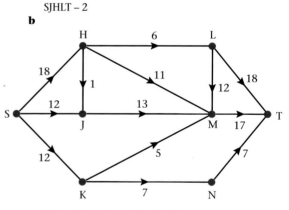

Value of maximum flow is 42

4 a

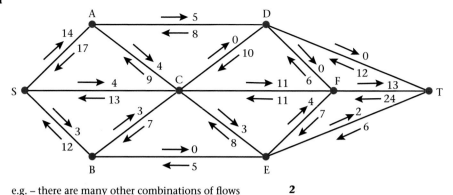

e.g. – there are many other combinations of flows possible
SACFT – 4
SADCFT – 5
SCFT – 2
SCEFT – 2
SBCET – 1

b

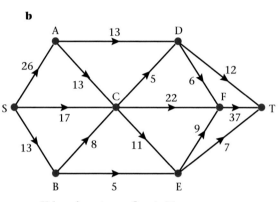

Value of maximum flow is 56

2

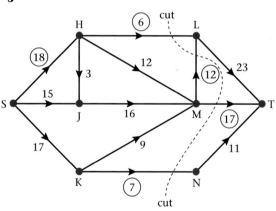

There are 2 cuts of value 38, both shown on the diagram. Flow is 38.
By the maximum flow–minimum cut theorem flow is maximum.

Exercise 6E

1

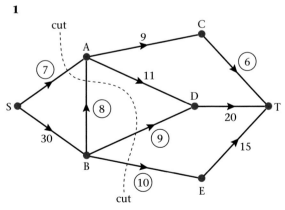

capacity of cut = 34
current flow = 34
so by maximum flow–minimum cut theorem flow is maximal

3

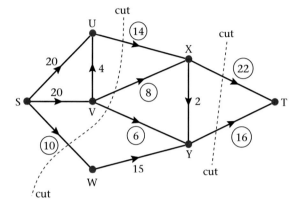

capacity of cut = 42
current flow = 42
so by maximum flow–minimum cut theorem flow is maximum

4

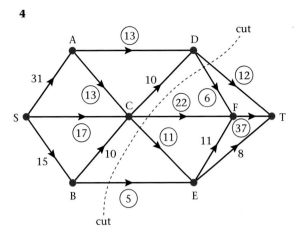

value of cut = 56
value of flow = 56
by maximum flow–minimum cut theorem flow is
maximum

Mixed exercise 6F

1 a

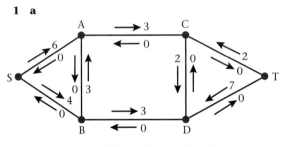

e.g. – many different flow combinations
SBDT – 3
SACT – 2
SBACDT – 1
Value of flow = 6

b Cut through AC and BD minimum cut = 6 so by
maximum flow = minimum cut theorem, flow is
maximum

2 a Applied   Idea of flow through a system, idea of
directed flow.
e.g. traffic moving through a one-way
system of roads.

b

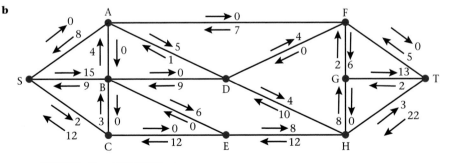

e.g.   SBEHGT – 6 and SBADFGT – 4
or   SBADHGT – 4 and SCBEHT – 2 and SBEHGT – 4
etc.

c e.g.

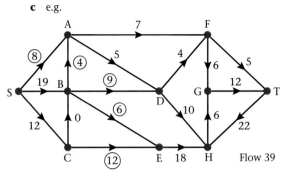

Flow 39

d Cut through SA, BA, BD, BE and CE
e The arcs are saturated.

3 a A, F, G and H, possible flow in > possible flow out
b e.g.

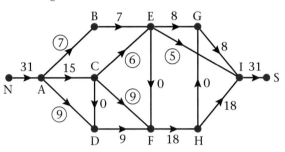

c Using labelling procedure
e.g.

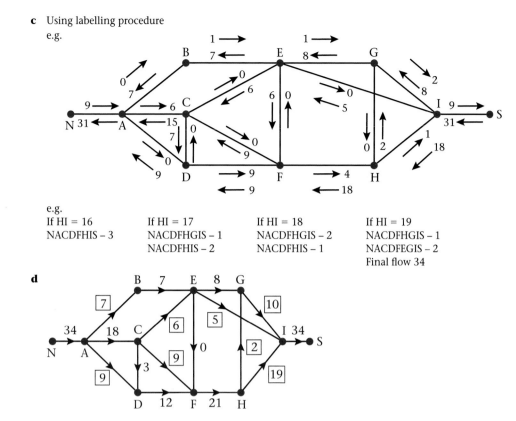

e.g.

If HI = 16	If HI = 17	If HI = 18	If HI = 19
NACDFHIS – 3	NACDFHGIS – 1	NACDFHGIS – 2	NACDFHGIS – 1
	NACDFHIS – 2	NACDFHIS – 1	NACDFEGIS – 2
			Final flow 34

d

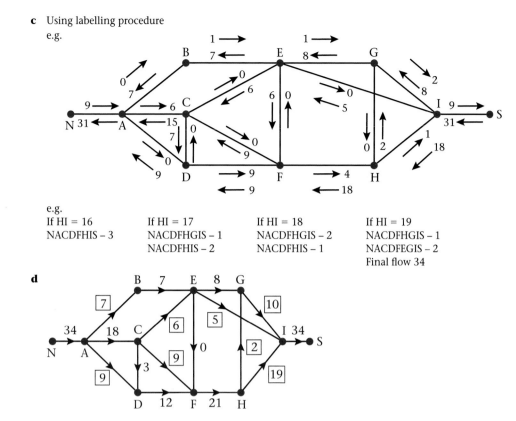

e and **f** cuts go through either G1, E1 and H1 *or* AB, CE, EF, HG and HI

4 a

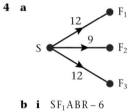

b i $SF_1ABR – 6$
ii $SF_3CR – 8$

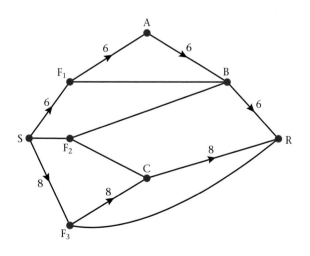

c i

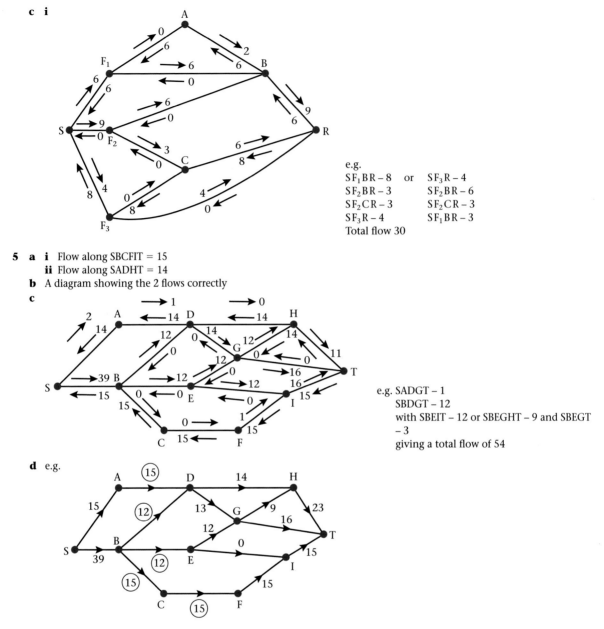

e.g.
$SF_1BR - 8$ or $SF_3R - 4$
$SF_2BR - 3$ $SF_2BR - 6$
$SF_2CR - 3$ $SF_2CR - 3$
$SF_3R - 4$ $SF_1BR - 3$
Total flow 30

5 a i Flow along SBCFIT = 15
   ii Flow along SADHT = 14
 b A diagram showing the 2 flows correctly
 c

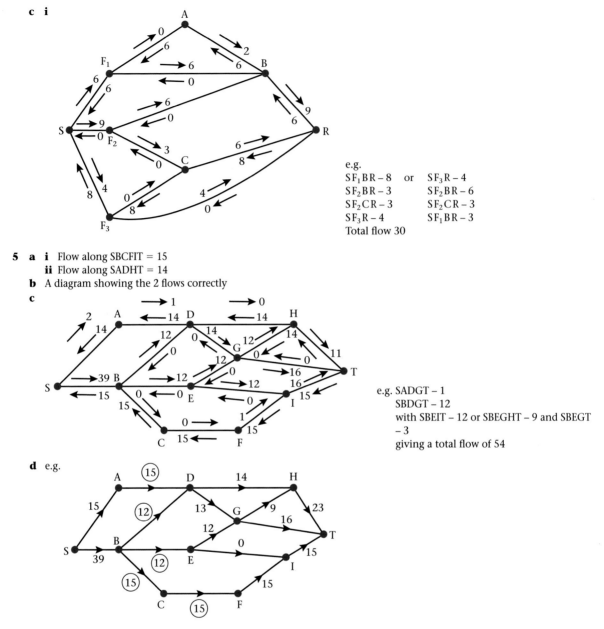

e.g. SADGT – 1
    SBDGT – 12
    with SBEIT – 12 or SBEGHT – 9 and SBEGT – 3
    giving a total flow of 54

 d e.g.

f The flow into D and into C could not increase.
so increase the flow along BE

6 a $v = 7$, $w = 6$, $x = 8$, $y = 3$, $z = 11$ (conservation of flow)
 b

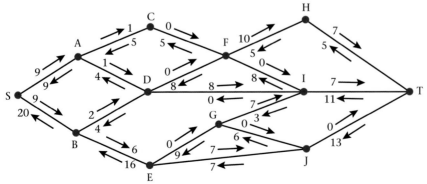

Increasing flow by an additional 3
e.g. SBDIT – 2
       SADIT – 1
Additional flow increases (reversing initial flow)
e.g. SBEJGIT – 4
       SBEJGIFHT – 2
Flow up to maximum – 38

c e.g.

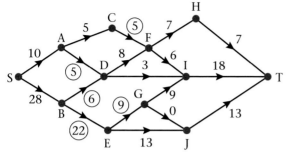

d i 12 + 18 + 13 = 43
  ii CF, AD, BD, BE
  iii max flow – min cut theorem
       e.g. The minimum cut separates the source
             from the sink. Any additional flow must
             cross this cut at some point. Since the
             arcs in the minimum cut are saturated no
             additional flow can be transported along
             these arcs.
             Hence no additional flow is possible.

7 a c_1 – 40 c_2 – 56
  b max flow = min cut = 40
  c e.g. Flow into F is 16 ∴ flow into G is 24. The
             flow along DG is 8
             ∴ Flow along GT is 24
  d e.g. Flow into A = flow out of A ∴ flow along AD
             $\leqslant 12$
             Flow into D = flow out of F = 21
             so flow along AD + flow along BD = 21
             ∴ flow along AD and BD could be 12 + 9 or
             11 + 10
             ∴ possible flows are 20 and 19
  e SA = 20

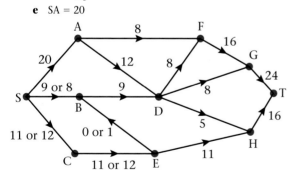

SA = 19

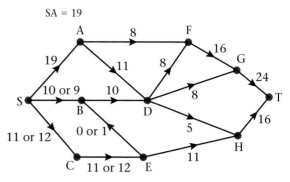

f There are 2 more – CE could be 11 or 12 in each
     case.

Exercise 7A

1 a Shortest route SBDHT length 87
  b Longest route SAEGT length 100

2 a Shortest route length is 88 with route SACGT
  b Longest route length is 97 with route SBFHT

3 a Shortest route length is 93 with route SBEHT
  b Longest route length is 101 with route SCFJT

4 a Shortest route length is 84 with route SBFGT
  b Longest route length is 95 with route SBDHT

5 The maximum profit is £145 000
     The maximum route is SDFT
     In practical terms the company's strategy is:
     Year 1 – advertise in both TV and Radio
     Year 2 – advertise on TV only.

Exercise 7B

1 a Minimax route SCDHT or SCEHT – both of value 30
  b Maximin route SCEGT of value 20

2 a Minimax route SBEHT of value 33
  b Maximin route SAEIT of value 14

3 a Minimax route SBEIT or SBFIT – both of value 27
  b Maximin routes SAEIT, SBEIT, SBFIT, SCEIT, SCFIT
       all of value 18

4 a Minimax route SBFGT of value 27
  b Maximin routes SADGT and SBDGT both of value
       18

Exercise 7C

1 a Stage – time, in years, remaining
       State – resorts already created
       Action – resort to be opened
  b The minimax route is CAB with a value of £5500
       The order in which the results should be built is C
       then A then B

2 a Stage – phase being considered
       State – number of days remaining
       Action – number of days allocated
       Destination – number of days remaining
       Value – total costs

b The minimum cost is £66 000. The time should be allocated as follows:

Activity	Clearance	Repairing	Modernisation	Decorating
Number of days	5	15	5	5

3 The minimum cost is £210 000. The aircraft should be built as follows:

Month	March	April	May	June
Number of aircraft built in each month	3	0	3	2

4 The minimum route is:
Home – A – E – F – I – Home
with a value of £2700

Mixed exercise 7D

1 Minimax route is SCEGT value 22

2 The maximin route is SBDHT of value 21

3 **a** and **b** There are two possible courses of action each of value £65

Product	Butter	Cheese	Yoghurt
Units to be used	2	2	1

Product	Butter	Cheese	Yoghurt
Units to be used	2	3	0

4 Tracing back there are two routes
SC, CF, FT ⇒ SCFT
SA, AD, DT ar SADT
Maximum altitude on these routes is 40 ($\times$100 ft) = 4000 ft

5 **a**

Stage	State	Action	Cost	Total cost
2	0	A	2	2
		C	3	3 *
	1	A	2	2
		B	3	3
		C	6	6 *
	2	A	1	1
		B	2	2 *
1	0	A	2	2 + 3 = 5
		C	3	3 + 6 = 9 *
	1	A	1	1 + 3 = 4
		B	3	3 + 6 = 9 *
		C	6	6 + 2 = 8
	2	A	5	5 + 6 = 11 *
		B	5	5 + 2 = 7
0	0	A	4	4 + 9 = 13
		B	3	3 + 9 = 12
		C	5	5 + 11 = 16 *

b Hence maximum profit is 16
Tracing back through calculations the optimal strategy is CAC

Review exercise 2

1 **a** A game in which the gain to one player is equal to the loss of the other.

b If there is a stable *solution* in a game, the *location* of this stable solution is called the saddle point.
It is the point where row maximin = column minimax

2 **a** $C_1 = 7 + 14 + 0 + 14 = 35$
$C_2 = 7 + 14 + 5 = 26$
$C_3 = 8 + 9 + 6 + 8 = 31$

b 26

c Using EJ (capacity 5) e.g. – will increase flow by 1 – i.e. increase it to 27 since only one more unit can leave E.
– BEJL – 1
Using FH (capacity 3) e.g. – will increase flow by 2 – i.e. increase it to 28 since only two more units can leave F.
– BFHJL – 2
Thus choose option 2 add FH capacity 3.

3 a

	A(I)	A(II)
B(I)	3	−4
B(II)	−2	1
B(III)	−5	4

b Let q_1 be the probability that B plays row 1
Let q_2 be the probability that B plays row 2
Let q_3 be the probability that B plays row 3
Let value of the game be v and let $V = v + 6$
where $q_1, q_2, q_3 \geqslant 0$
e.g. maximise $P = V$
subject to $V - 9q - 4q_2 - q_3 + r = 0$
$\qquad\qquad V - 2q_1 - 7q_2 - 10q_3 + s = 0$
$\qquad\qquad q_1 + q_2 + q_3 + t = 1$

4

Month	May	June	July	August	September
Production schedule	4	4	5	5	4

Cost £2300

5 a $x = 9, y = 16$

b Initial flow = 53

c

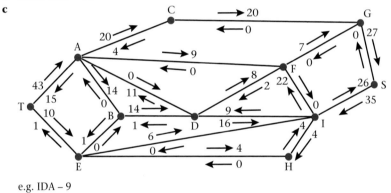

e.g. IDA – 9
IFDA – 24
max flow – 64

d

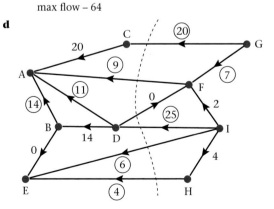

6 a Play safe is A plays II or IV and B plays III

c Value of game to B is $-(-1) = 1$

7 a

Stage	Initial state	Action	Destination	Value
	D	DT	T	8*
1	E	ET	T	10*
	F	FT	T	6*
	A	AD	D	max(7, 8) = 8*
		AE	E	max(8, 10) = 10
2	B	BE	E	max(9, 10) = 10
		BF	F	max(3, 6) = 6*
	C	CE	E	max(6, 10) = 10
		CF	F	max(9, 6) = 9*
		SA	A	max(9, 8) = 9
3	S	SB	B	max(7, 6) = 7*
		SC	C	max(6, 9) = 9

b Minimax route is SBFT

Maximum amount of fuel used is 7 units

8 a

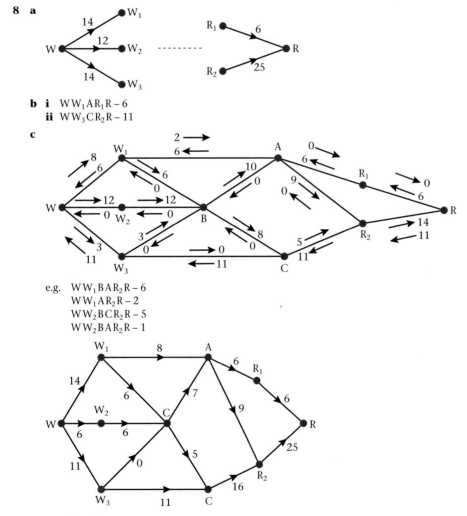

b i $WW_1AR_1R - 6$

 ii $WW_3CR_2R - 11$

c

e.g. $WW_1BAR_2R - 6$
$WW_1AR_2R - 2$
$WW_2BCR_2R - 5$
$WW_2BAR_2R - 1$

max flow 31

d 12 – For this network, but may be different for others

e Make no use of the opportunity.

All arcs out of A and C are saturated, so the total flow cannot be increased when the number of van loads from A or C to R_1 or R_2 is increased.

9 **b** Emma should play R_1 with probability $\frac{4}{11}$

R_2 with probability $\frac{7}{11}$

The value of the game is $-\frac{2}{11}$ to Emma.

c Value to Freddie $\frac{2}{11}$, matrix $\begin{pmatrix} 4 & -2 \\ 1 & -1 \\ -3 & 2 \end{pmatrix}$

10 **a** Adds S and T and arcs.

$SS_1 \geqslant 45$, $SS_2 \geqslant 35$, $T_1T \geqslant 24$, $T_2T \geqslant 58$

b Using conservation of flow through vertices $x = 16$ and $y = 7$

c $C_1 = 86$, $C_2 = 81$

d

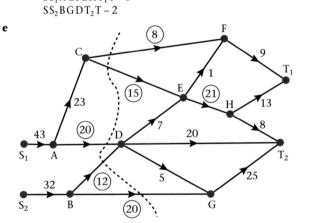

e.g. $SS_1ADEHT_2T - 2$

$SS_1ACFEHT_1T - 3$

$SS_2BGDT_2T - 2$

e

Flow 75

11 a The route from start to finish in which the arc of minimum length is as large as possible.
Example must be practical, involve choice of route, have arc 'costs'.

b

Stage	State	Action	Value
1	H	HK	18*
	I	IK	19*
	J	JK	21*
2	F	FH	min (16, 18) = 16
		FI	min (23, 19) = 19*
		FJ	min (17, 21) = 17
	G	GH	min (20, 18) = 18
		GI	min (15, 19) = 15
		GJ	min (28, 21) = 21*
3	B	BG	min (18, 21) = 18*
	C	CF	min (25, 19) = 19*
		CG	min (16, 21) = 16
	D	DF	min (22, 19) = 19*
		DG	min (19, 21) = 19*
	E	EF	min (14, 19) = 14*
4	A	AB	min (24, 18) = 18
		AC	min (25, 19) = 19*
		AD	min (27, 19) = 19*
		AE	min (23, 14) = 14

c Routes ACFIK, ADFIK, ADGJK

12 a $x = 3, y = 26$

b

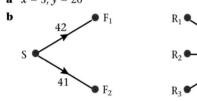

c

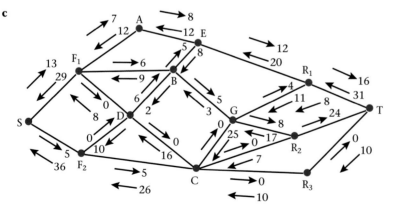

e.g. $SF_1AER_1T - 7$
  $SF_1BER_1T - 5$
  $SF_1BGR_1T - 1$
  $SF_2CBDGR_2T - 4$

d e.g.

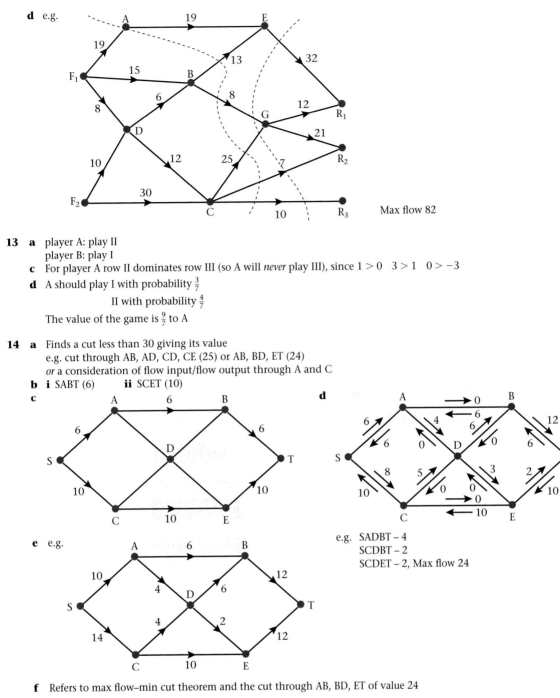

Max flow 82

13 a player A: play II
player B: play I

c For player A row II dominates row III (so A will *never* play III), since $1 > 0$ $3 > 1$ $0 > -3$

d A should play I with probability $\frac{3}{7}$
II with probability $\frac{4}{7}$
The value of the game is $\frac{9}{7}$ to A

14 a Finds a cut less than 30 giving its value
e.g. cut through AB, AD, CD, CE (25) or AB, BD, ET (24)
or a consideration of flow input/flow output through A and C

b i SABT (6)   **ii** SCET (10)

c

d

e.g. SADBT – 4
SCDBT – 2
SCDET – 2, Max flow 24

e e.g.

f Refers to max flow–min cut theorem and the cut through AB, BD, ET of value 24

15 a Row 1 dominates row 2 so A will never choose R2
Column 1 dominates column 3 so B will never choose C3
Thus Row 2 and column 3 may be deleted.

b A should play row 1 with probability $\frac{3}{5}$
A should play row 2 never
A should play row 3 with probability $\frac{2}{5}$
B should play column 1 with probability $\frac{2}{5}$
and column 2 with probability $\frac{3}{5}$
and column 3 never.
Value of game is $4\frac{1}{5}$ to A

16 a i maximum flow along SACDT = 50
ii maximum flow along SBT = 100

b

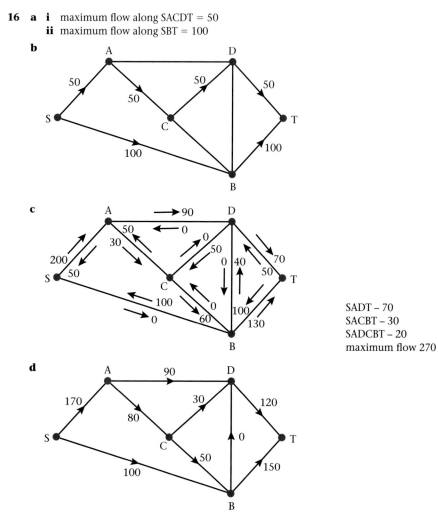

c

SADT – 70
SACBT – 30
SADCBT – 20
maximum flow 270

d

17 a £630

b

Month	August	September	October	November
Make	3	4	4	2

Cost = £1540

c Profit per cycle = 13 × 1400   cost of Kris' time   = £2000
                    = 18 200   cost of production = £1540
∴ Total profit = 18 200 − 3540
             = £14 660

18 a A, E and G
b 45
c

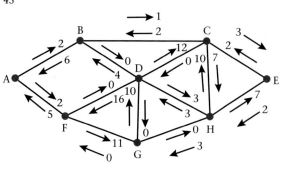

e.g.   EHD – 2
       ECHD – 1

d

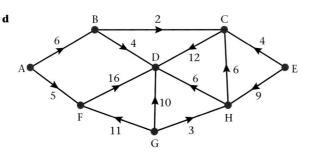

Maximum flow 48

19 b Row 2 dominates Row 3

Column 1 dominates column 4

c Let A play row R_1 with probability p_1, R_2 with probability p_2 and R_3 with probability p_3

e.g.   maximise $P = V$

subject to $V - p_1 - 2p_2 - 4p_3 \leqslant 0$

                  $V - 4p_1 - 6p_2 - p_3 \leqslant 0$

                  $V - 6p_1 - 5p_2 - 2p_3 \leqslant 0$

                  $p_1 + p_2 + p_3 \leqslant 1$

                  $V, p_1, p_2, p_3 \geqslant 0$

20 a A zero-sum game is one in which the sum of the gains for all players is zero. (o.e.)

c A should play I $\frac{1}{3}$ of time and II $\frac{2}{3}$ of time; value (to A) $= 3\frac{2}{3}$

d Let B play I with probability q_1, II with probability q_2 and III with probability q_3

e.g. $\begin{bmatrix} -5 & -3 \\ -2 & -5 \\ -3 & -4 \end{bmatrix} \rightarrow \begin{bmatrix} 1 & 3 \\ 4 & 1 \\ 3 & 2 \end{bmatrix}$

Maximise $P = V$

Subject to $V - q_1 - 4q_2 - 3q_3 \leqslant 0$

            $V - 3q_1 - q_2 - 2q_3 \leqslant 0$ $q_1 + q_2 + q_3 \leqslant 1$

            $V, q_1, q_2, q_3 \geqslant 0$

21 a Stage – number of weeks to finish

State – show being attended

Action – next journey to undertake

b

Stage	State	Action	Value
1	F	F – Home	$500 - 80 = 420\star$
	G	G – Home	$700 - 90 = 610\star$
	H	H – Home	$600 - 70 = 530\star$
2	D	DF	$1500 - 200 + 420 = 1720$
		DG	$1500 - 160 + 610 = 1950\star$
		DH	$1500 - 120 + 530 = 1910$
	E	EF	$1300 - 170 + 420 = 1550$
		EG	$1300 - 100 + 610 = 1810\star$
		EH	$1300 - 110 + 530 = 1720$
3	A	AD	$900 - 180 + 1950 = 2670\star$
		AE	$900 - 150 + 1810 = 2560$
	B	BD	$800 - 140 + 1950 = 2610\star$
		BE	$800 - 120 + 1810 = 2490$
	C	CD	$1000 - 200 + 1950 = 2750\star$
		CE	$1000 - 210 + 1810 = 2600$
4	Home	Home – A	$-70 + 2670 = 2600\star$
		Home – B	$-80 + 2610 = 2530$
		Home – C	$-150 + 2750 = 2600\star$

c

Home — A — D — G
Home — C — D — G

Total profit £2600

22 a

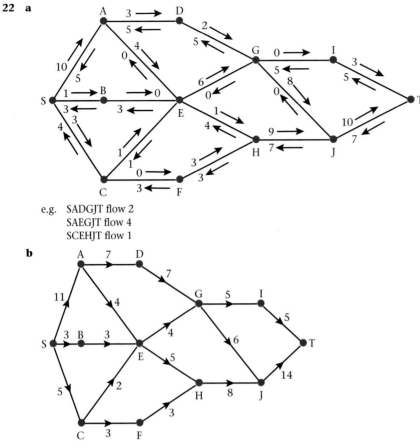

e.g. SADGJT flow 2
SAEGJT flow 4
SCEHJT flow 1

b

c 19 students

Examination style paper

1 A – S (33)
B – Q (34)
C – R (34)   134 seedlings
D – P (33)

2 a $x = 22$; $y = 31$
b Saturated arcs SA; SB; AD; BA; BC; CE; CF; ET; FT
c Flow value is 51
d Capacity of $C_1 = 22 + 8 + 12 + 3 + 14 = 59$ Capacity of $C_2 = 17 + 31 + 4 + 5 = 57$
e Flow-augmenting route: $SCBEDT - 3$
f Cut through SA, BA, BE, CE, CF

3 a

	A	B	C	D	E	F	G
A	–	26	56	24	39	37	53
B	26	–	51	19	21	46	39
C	56	51	–	32	47	28	48
D	24	19	32	–	15	40	29
E	39	21	47	15	–	25	18
F	37	46	28	40	25	–	20
G	53	39	48	29	18	20	–

b A   D   E   G   F   C   B   A $= 182$
      24 15 18 20 28 51 26

c $E_{15} D_{19} B_{26} A_{37} F_{20} G_{48} C_{47} E = 212$

d 182 is the better upper bound since it is lower.

e 147

f $147 <$ lenth of optimal route $\leqslant 172$

4 a

	P	Q	R	
A	11			11
B	19	9		28
C		31	5	36
D			25	25
	30	40	30	100

b Improved solution:

	P	Q	R	
A	2		9	11
B	28			28
C		31	5	36
D			25	25
	30	40	30	100

c £2780

d The current solution is not optimal, there is a negative improvement index.

5

Stage	State	Action	Destination	Value
1	H	HJ IJ	J J	37* 36*
2	E	EH EI	H I	max (36, 37) = 37* max (40, 36) = 40
	F	FH FI	H I	max (38, 37) = 38* max (39, 36) = 39
	G	GH GI	H I	max (36, 37) = 37* max (39, 36) = 39
3	B	BE BF BG	E F G	max (37, 37) = 37* max (36, 38) = 38 max (41, 37) = 41
	C	CE CF CG	E F G	max (38, 37) = 38* max (39, 38) = 39 max (38, 37) = 38*
	D	DE DF DG	E F G	max (40, 37) = 40 max (41, 38) = 41 max (36, 37) = 37*
4	A	AB AC AD	A C D	max (38, 37) = 38 max (37, 39) = 39 max (37, 37) = 37*

Minimax route is ADGHT value 37

6 Greg should play 1 with probability $\frac{3}{8}$ and play 2 with probability $\frac{5}{8}$. The value to Greg is $-\frac{1}{8}$

7 a

b.v.	x	y	z	r	s	t	Value	θ values
r	2	1	④	1	0	0	20	$\frac{20}{4} = 5 \leftarrow$
s	6	3	2	0	1	0	15	$\frac{15}{2} = 7.5$
t	1	4	-2	0	0	1	8	negative value
P	-3	-1	-4	0	0	0	0	

b.v.	x	y	z	r	s	t	Value	Row operations	θ values
z	$\frac{1}{2}$	$\frac{1}{4}$	1	$\frac{1}{4}$	0	0	5	R1 ÷ 4	$\frac{5}{\frac{1}{2}} = 10$
s	⑤	$\frac{5}{2}$	0	$-\frac{1}{2}$	1	0	5	R2 − 2R1	$\frac{5}{5} = 1 \leftarrow$
t	2	$\frac{9}{2}$	0	$\frac{1}{2}$	0	1	18	R3 + 2R1	$\frac{18}{2} = 9$
P	-1	0	0	1	0	0	20	R4 + 4R1	

b.v.	x	y	z	r	s	t	Value	Row operations
z	0	0	1	$\frac{3}{10}$	$-\frac{1}{10}$	0	$\frac{9}{2}$	R1 − $\frac{1}{2}$R2
x	1	$\frac{1}{2}$	0	$-\frac{1}{10}$	$\frac{1}{5}$	0	1	R2 ÷ 5
t	2	$\frac{7}{2}$	0	$\frac{7}{10}$	$-\frac{2}{5}$	1	16	R3 − 2R2
P	0	$\frac{1}{2}$	0	$\frac{9}{10}$	$\frac{1}{5}$	0	21	R4 + R2

$P = 21$ $x = 1$ $y = 0$ $z = \frac{9}{2}$ $r = 0$ $s = 0$ $t = 16$

b This tableau is optimal, there are no negative numbers in the profit row.

8 Let p_1 be the probability that A plays 1
Let p_2 be the probability that A plays 2
Let p_3 be the probability that A plays 3
where $p_1, p_2, p_3 \geqslant 0$

Let $v =$ the value of the original game to player A
Then $V = v + 3 =$ value of the new game to player A

Maximise $P = V$ so $P - V = 0$
Subject to:
If B plays column 1: $6p_1 + 7p_2 + 3p_3 \geqslant V$
If B plays column 2: $4p_1 + 2p_2 + 5p_3 \geqslant V$
If B plays column 3: $p_1 + 3p_2 + 8p_3 \geqslant V$
where $p_1, p_2, p_3 \geqslant 0$

Index